No. 3.

Papers
relating to the duties of the Corps of
Topographical Engineers.

TO COL. J. J. ABERT,
Chief Corps of Topographical Engineers.

Sir : I have endeavored, in the following pages, to comply with your instructions by presenting, in as condensed a form as practicable, such Tables and Formulæ as may prove most useful to an officer engaged in the active duties of a survey.

In the selection of the matter it has been my aim to present the best methods, as far as they have been practised by us, or may be applicable to the nature of our duties, in such forms as to be convenient for reference, and still secure a high degree of accuracy in the reduction of such observations as may be requisite for the minute survey of a limited extent of country, as well as for the exact determination of Geographical Positions or for distant Explorations.

With such a subject I can lay claim to but little that is original, and although aware of the many imperfections in this Collection, I still trust that it may not be without its utility, and that as a *Manual* of easy reference it may meet the wants of my brother officers.

Although every precaution has been taken to ensure accuracy of print, it is not improbable that some errors may have escaped correction. A table of errata is appended, and I would be obliged by the communication of any others that may be detected.

Very respectfully,

Your obedient servant,

THOMAS J. LEE,
Cap. Top. Engineers

Washington, *August* 8, 1849.

ERRATA.

Page 9; 27878400 square feet = 1 square mile.

Page 64; $Dp = (5.0857556) \cos \varphi - (2.00835) \cos 3\varphi + $ etc. ;

Page 145; opposite 24 hours, read $3^m 56^s.555$.

PREFATORY.

The following explanations of the sources from which the several portions of this Collection were derived, will serve to establish the degree of confidence with which each may be received.

Part I.

Pp. 1—3.   Baily, Astronomical Tables and Formulæ.
       Francœur, Géodésie, Paris, 1840.

4.   Francœur, Géodésie.

5—7.   Bégat, Traité de Géodésie, Paris, 1839.

8—14.   Weights and measures. Those of the United States will be found in the Report of Professor Bache, Superintendent of Weights and Measures, July 30, 1848, Ex. Doc., No. 84, 30th Congress, 1st session. The remaining part of the article is from Brande's Dictionary of Science and Art, American edition, the quantities having been compared, when practicable, with Alexander's Universal Dictionary of Weights and Measures, Baltimore, 1850, and with the tables in Appleton's Dictionary of Mechanics and Engineering.

14—15.   These tables are from the Edinburg Philosophical Journal of October, 1837.

16—17.   Abridged from tables in Hulsse's Sammlung Mathematischer Tafeln, Leipsic, 1840, with the addition of the relation of Spanish and Mexican measures.

18.   Claudel, Aide Memoire des Ingenieurs, etc., Paris, 1849. The length of the Spanish vara was compared with its value given in Francœur, Bégat, Brande, and Hulsse's works.

19.   Ordnance Manuel, 1850.

20—22.   Bégat, Géodésie.

Pp.   23.   McNiel's Railway Tables, London, 1833. Beardmore, Hydraulic and other Tables, London, 1852.

24—25. Storrow on Water Works, Boston, 1833.

26—30. Railroad Manual, by Brevet Lieut. Col. Long, Corps Topographical Engineers, Baltimore, 1829.

31—32. Davies' Surveying.

33—47. Abridged from the Traverse Tables of Captain J. T. Boileau, Bengal Engineers.

48—49. Beardmore, Hydraulic and other tables, London, 1852.

50. Regulations of the Subsistence Department.

Part II.

53. Bégat Géodésie.

54—55. Galbraith, Mathematical Tables and Formulæ, Edinburg, 1834.

56. Francœur, Géodésie.

57. Baily, Astronomical Tables and Formulæ, London, 1827.

58—59. Galbraith, Mathematical Tables, etc.

60. Baily, Astronomical Tables.—Bégat, Géodésie.

61. These values were carefully compared with the original in the Astronomische Nachrichten, No. 438.

62—64. I have adopted the yard as a unit, it being the unit of our lineal measures, and have, in the text, given the reasons for adhering to Kater's value of the metre. I have also preferred the *established* ratio of the metre to the toise, to that derived from Mr. Hassler's comparisons. (See Hassler's report of 1832, Doc. No. 299, 22d Congress, 1st session.)

In reducing Bessel's Terrestrial Elements to English yards, and in other computations hereafter to be noticed, I was fortunate in securing the services of Mr. John Downes, now attached to the American Nautical Almanac establishment, whose well known reputation is the surest evidence of their accuracy. These reductions were also compared with a computation of my own.

B

Part III.

T. J. L.

CONTENTS.

Part III.—*Astronomy.*

c

TABLES AND FORMULÆ.

PART I.

MISCELLANEOUS.

I. *Equivalent Expressions.*

$$\text{Sin }^2x + \cos {}^2x = 1.$$

$$\text{Sin } x = \cos x \,.\, \text{tang } x$$

$$= \frac{\cos x}{\cot x}$$

$$= \sqrt{1 - \cos {}^2x}$$

$$= 2 \sin \tfrac{1}{2} x \,.\, \cos \tfrac{1}{2} x$$

$$= \frac{\text{tang } x}{\sqrt{1 + \text{tang }^2x}}$$

$$= \frac{1}{\text{cosecant } x}$$

$$\text{Cos } x = \frac{\sin x}{\text{tang } x}$$

$$= \sin x \,.\, \cot x$$

$$= \sqrt{1 - \sin {}^2x}$$

$$= \cos {}^2\tfrac{1}{2}x - \sin {}^2\tfrac{1}{2} x$$

$$= \frac{1}{\text{secant } x}$$

$$\text{Tang } x = \frac{\sin x}{\cos x}$$

$$= \frac{1}{\cot x}$$

$$= \frac{\sin x}{\sqrt{1 - \sin {}^2x}}$$

$$= \frac{\sin 2 x}{1 + \cos 2 x}$$

$$\text{Cotang } x = \frac{1}{\text{tang } x}$$

$$\text{Secant } x = \frac{1}{\cos x}$$

$$\text{Cosecant } x = \frac{1}{\sin x}$$

$$\text{Versed sin } x = 1 - \cos x = 2 \sin^2 \tfrac{1}{2} x$$

$$\text{Co-versed sin } x = 1 - \sin x$$

$$\text{Chord } x = 2 \sin \tfrac{1}{2} x$$

$$\text{Sin} (A \pm B) = \sin A \cos B \pm \sin B \cos A$$

$$\text{Cos} (A \pm B) = \cos A \cos B \mp \sin A \sin B$$

$$\text{Sin } 2A = 2 \sin A \cos A$$

$$\text{Cos } 2A = 2 \cos^2 A - 1 = 1 - 2 \sin^2 A$$
$$= \cos^2 A - \sin^2 A$$

$$2 \cos^2 \tfrac{1}{2} A = 1 + \cos A$$

$$2 \sin^2 \tfrac{1}{2} A = 1 - \cos A$$

$$\text{Tang} (A \pm B) = \frac{\text{tang } A \pm \text{tang } B}{1 \mp \text{tang } A \text{ tang } B}$$

$$\text{Tang } \tfrac{1}{2} A = \sqrt{\frac{1 - \cos A}{1 + \cos A}} = \frac{1 - \cos A}{\sin A}$$

$$\text{Sin } A \pm \sin B = 2 \sin \tfrac{1}{2} (A \pm B) \cos \tfrac{1}{2} (A \mp B)$$

$$\text{Cos } A + \cos B = 2 \cos \tfrac{1}{2} (A + B) \cos \tfrac{1}{2} (A - B)$$

$$\text{Cos } A - \cos B = 2 \sin \tfrac{1}{2} (A + B) \sin \tfrac{1}{2} (B - A)$$

$$\text{Sin}^2 A - \sin^2 B = \sin (A + B) \sin (A - B)$$

$$\text{Cos}^2 A - \sin^2 B = \cos (A + B) \cos (A - B)$$

$$\text{Tang } A \pm \text{tang } B = \frac{\sin (A \pm B)}{\cos A \cos B}$$

$$\text{Cot } A \pm \cot B = \frac{\sin (A \pm B)}{\sin A \sin B}$$

$$\frac{\text{Sin A} + \sin \text{B}}{\text{Sin A} - \sin \text{B}} = \frac{\tan \frac{1}{2} (\text{A} + \text{B})}{\tan \frac{1}{2} (\text{A} - \text{B})}$$

$$\frac{1 \pm \sin \text{A}}{1 \mp \sin \text{A}} = \tan^2 (45° \pm \tfrac{1}{2} \text{A})$$

$$\frac{1 \pm \sin \text{A}}{\text{Cos A}} = \tan (45° \pm \tfrac{1}{2} \text{A})$$

II. *Trigonometrical Series.*

$$\text{Sin A} = \text{A} - \frac{\text{A}^3}{2.3} + \frac{\text{A}^5}{2.3.4.5} - \frac{\text{A}^7}{2.3.....7} + \text{etc.}$$

$$\text{Cos A} = 1 - \frac{\text{A}^2}{2} + \frac{\text{A}^4}{2.3.4} - \frac{\text{A}^6}{2.....6} + \text{etc.}$$

$$\text{Arc A} = \sin \text{A} + \frac{\sin^3 \text{A}}{2.3} + \frac{3 \sin^5 \text{A}}{2.4.5} + \frac{3.5 \sin^7 \text{A}}{2.4.6.7} \text{ etc.}$$

$$= \tan \text{A} - \tfrac{1}{3} \tan^3 \text{A} + \tfrac{1}{5} \tan^5 \text{A} - \tfrac{1}{7} \tan^7 \text{A}..$$

$$\text{Log sin A} = \log \text{A} + \log \left(1 - \frac{x^2}{6} + \frac{x^4}{120} - \text{etc.}\right)$$

$$= \log \text{A} - \text{M} \left(\frac{x^2}{6} + \frac{x^4}{180} + \frac{x^4}{2835}\right)$$

$$\text{M} = \text{logarithmic modulus} = 0.4342945.....$$

$$\text{Log M} = 9.6377843113....$$

III. *Table of signs of Trigonometrical lines.*

Quadrants.	Sine.	Cosine.	Tang.	Cot.	Secant.	Cosecant.
1. 5. 9.	+	+	+	+	+	+
2. 6. 10.	+	—	—	—	—	+
3. 7. 11.	—	—	+	+	—	—
4. 8. 12, &c.	—	+	—	—	+	—

IV. *Ratio of the circumference of a circle to its diameter.*

$$\pi = 3.14159\ 26535\ 898\ldots\ldots$$
$$\text{Log } \pi = 0.49714\ 98726\ 941\ldots\ldots$$

The radius being unity, the number of degrees in an arc equal to radius $= r^{\circ} = \dfrac{180^{\circ}}{\pi.} = \dfrac{1}{\text{arc }1^{\circ}} = 57^{\circ}.29578 = 57^{\circ}.17'.44''.8.$

The number of minutes $= r' = \dfrac{10800'}{\pi} = \dfrac{1}{\text{arc }1'}$ or $\dfrac{1}{\sin 1'} = 3437'.74677.$

The number of seconds $= r'' = \dfrac{648000''}{\pi} = \dfrac{1}{\sin 1''} = 206264''.80625.$

$$\text{Log } r^{\circ} = 1.75812\ 26324\ 09172$$
$$\text{Comp log } r^{\circ} = 8.24187\ 73675\ 90828$$
$$\text{Log } r' = 3.53627\ 38827\ 92816$$
$$\text{Comp log } r' = 6.46372\ 61172\ 07184 = \text{log sin } 1'$$
$$\text{Log } r'' = 5.31442\ 51331\ 76459$$
$$\text{Comp log } r'' = 4.68557\ 48668\ 23541 = \text{log sin } 1''$$

Let a be the length of an arc of a circle whose radius is 1, and a'' the number of seconds in that arc, as

$$r'' = \frac{1}{\sin 1''} \text{ and } R : r'' :: a : a'' \text{ or } a'' = r'' a;\ a = a'' \sin 1''$$

In an equation, therefore, any arc a of a circle whose radius is 1, is expressed in seconds by changing a into $a'' \sin 1''$.

V. *Solution of Plane Triangles.*

In the following formulæ A, B, C, represent the angles, and a, b, c, the sides opposite, respectively.

1. Any plane triangle,

$$a^2 = b^2 + c^2 - 2\,b\,c\,\cos A$$

$$\frac{\sin A}{a} = \frac{\sin B}{b} = \frac{\sin C}{c}$$

$$\frac{\text{tang } \frac{1}{2}\,(A + B)}{\text{tang } \frac{1}{2}\,(A - B)} = \frac{\cot \frac{1}{2}\,C}{\text{tang } \frac{1}{2}\,(A - B)} = \frac{a + b}{a - b}$$

$$\sin \tfrac{1}{2}\,A = \left\{ \frac{(s - b)\,(s - c)}{b\,c} \right\}^{\frac{1}{2}}$$

$$\cos \tfrac{1}{2}\,A = \left\{ \frac{s\,(s - a)}{b\,c} \right\}^{\frac{1}{2}}$$

$$s = \frac{a + b + c}{2}$$

2. Right angled triangles,

making A $= 90°$ in the preceding, they become

$$a^2 = b^2 + c^2$$

$$b = a \sin B = a \cos C, \qquad c = a \sin C = a \cos B$$

$$\text{tang } B = \frac{b}{c} \qquad\qquad \text{tang } C = \frac{c}{b}$$

VI. *Solution of Spherical Triangles.*

a, b, c, represent the arcs, and A, B, C, the angles opposite.

1. Oblique spherical triangles

$$\frac{\sin A}{\sin a} = \frac{\sin B}{\sin b} = \frac{\sin C}{\sin c}$$

$$\begin{cases} \cos a = \dfrac{\cos b \, \sin (c + \phi)}{\sin \phi} \\ \cot \phi = \tang b \, \cos A \end{cases}$$

$$\begin{cases} \cos A = \dfrac{\cos B \, \sin (C - \phi)}{\sin \phi} \\ \cot \phi = \tang B \, \cos a \end{cases}$$

$$\begin{cases} \cot a \, \tang b = \dfrac{\sin (C + \phi)}{\sin \phi} \\ \cot \phi = \dfrac{\cot A}{\cos b} \end{cases}$$

Napier's Analogies.

$$\begin{cases} \tang \tfrac{1}{2} (a + b) = \tang \tfrac{1}{2} c \; \dfrac{\cos \frac{1}{2} (A - B)}{\cos \frac{1}{2} (A + B)} \\ \tang \tfrac{1}{2} (a - b) = \tang \tfrac{1}{2} c \; \dfrac{\sin \frac{1}{2} (A - B)}{\sin \frac{1}{2} (A + B)} \\ \tang \tfrac{1}{2} (A + B) = \cot \tfrac{1}{2} C \; \dfrac{\cos \frac{1}{2} (a - b)}{\cos \frac{1}{2} (a + b)} \\ \tang \tfrac{1}{2} (A - B) = \cot \tfrac{1}{2} C \; \dfrac{\sin \frac{1}{2} (a - b)}{\sin \frac{1}{2} (a + b)} \end{cases}$$

VI. *Solution of Spherical Triangles*—Continued.

$$\sin^2 \tfrac{1}{2}\, a = \frac{\sin S \cdot \sin (A - S)}{\sin B \cdot \sin C}$$

$$\cos^2 \tfrac{1}{2}\, a = \frac{\sin (B - S) \cdot \sin (C - S)}{\sin B \cdot \sin C}$$

$$\tan^2 \tfrac{1}{2}\, a = \frac{\sin S \cdot \sin (A - S)}{\sin (B - S) \cdot \sin (C - S)}$$

$$\sin^2 \tfrac{1}{2}\, A = \frac{\sin (s - b) \cdot \sin (s - c)}{\sin b \cdot \sin c}$$

$$\cos^2 \tfrac{1}{2}\, A = \frac{\sin s \cdot \sin (s - a)}{\sin b \cdot \sin c}$$

$$\tan^2 \tfrac{1}{2}\, A = \frac{\sin (s - b) \cdot \sin (s - c)}{\sin s \cdot \sin (s - a)}$$

In which S and s represent the half sum of the three angles diminished by 90° and the half sum of the three sides, respectively.

2. Right angled spherical triangles, a, being the hypothenuse.

$\cos a = \cos b \cdot \cos c$	$\cot B = \cot b \cdot \sin c$
$\cos a = \cot B \cdot \cot C$	$\cot C = \cot c \cdot \sin b$
$\cos B = \sin C \cdot \cos b$	$\tan b = \tan B \cdot \sin c$
$\cos C = \sin B \cdot \cos c$	$\tan c = \tan C \cdot \sin b$
$\tan b = \tan a \cdot \cos C$	$\sin b = \sin a \cdot \sin B$
$\tan c = \tan a \cdot \cos B$	$\sin c = \sin a \cdot \sin C$

I. *Weights and Measures of the United States.*

The *actual standard of length* is a brass scale of 82 inches in length, made by Troughton, of London, and now in the possession of the Treasury Department.

The *standard of weight* is the *troy pound*, copied in 1827, by Captain Kater, from the imperial troy pound of England, for the use of the Mint of the United States, and there deposited.

This pound is a standard at 30 inches of the Barometer and 62° of the Fahrenheit Thermometer.

The *units of capacity measure* are the *gallon* for *liquid* and the *bushel* for *dry* measure. The gallon is a vessel containing 58372.2 grains, (8.3389 pounds avoirdupois,) of the standard pound of distilled water, at the temperature of maximum density of water, the vessel being weighed in air in which the Barometer is 30 inches at 62° Fahrenheit. The bushel is a measure containing 543391.89 standard grains (77.6274 pounds avoirdupois) of distilled water, at the temperature of maximum density of water, and Barometer 30 inches at 62° Fahrenheit.

The gallon is thus the Wine gallon (of 231 cubic inches) nearly, and the bushel the Winchester bushel, nearly.

The temperature of maximum density of water was determined by Mr. Hassler to be 39°.83 Fahrenheit.

The avoirdupois pound is greater than the troy pound in the proportion of 7000 to 5760; that is, the avoirdupois pound is equivalent, in weight, to 7000 grains troy.

II. *English System of Measures.*

The unit of lineal measure is the *yard.* The yard is divided into 3 feet, and the foot subdivided into 12 inches. The multiples of the yard are the *pole* or *perch,* the *furlong,* and the *mile.* But the pole and furlong are now scarcely ever used, itinerary distances being reckoned in miles and yards. The following are the relations:

Inches.	Feet.	Yards.	Poles.	Furlongs.	Miles.
1	0.083	0.028	0.00505	0.00012626	0.0000157828
12	1.	0.333	0.06060	0.00151515	0.00018939
36	3.	1.	0.1818	0.004545	0.00056818
198	16.5	5.5	1.	0.025	0.003125
7920	660.	220.	40.	1.	0.125
63360	5280.	1760.	320.	8.	1.

Measures of Superficies.

In square measure the yard is subdivided as in general measure into *feet* and *inches;* 144 square inches being equal to a square foot. For land measure the multiples of the yard are the *pole*, the *rood*, and the *acre.* Very large surfaces, as of whole countries, are expressed in square miles.

The following are the relations of square measure:

Sq. feet.	Sq. yards.	Poles.	Roods.	Acres.	Sq. mile.
1.	0.1111	0.00367309	0.000091827	0.000022957	
9.	1.	0.0330579	0.000826448	0.000206612	
272.25	30.25	1.	0.025	0.00625	
10890.	1210.	40.	1.	0.25	
43560.	4840.	160.	4.	1.	
292800.	3097600.	102400.	2560.	640.	1.

Log. 3097600 = 6.4910253.

III. *Measures of Volume.*

Solids are measured by cubic yards, feet, and inches; 1728 cubic inches making a cubic foot, and 27 cubic feet a cubic yard. For all sorts of liquids, grain, and other dry goods, the standard measure is declared, by the act of 1824, to be the *imperial gallon*, the capacity of which is determined immediately by weight, and remotely by the standard of length, in the following manner: According to the act, the imperial standard gallon contains 10 pounds avoirdupois weight of distilled water, weighed in air at the temperature of 62° Fahrenheit's Thermometer, the Barometer being at 30 inches. The pound avoirdupois contains 7,000 troy grains; and it is declared that a cubic inch of distilled water (temperature 62°, barometer 30 inches) weighs 252.458 grains. Hence the contents of the imperial standard gallon are 277.274 cubic inches. The parts of the gallon are *quarts* and *pints*. Its multiples are the *peck*, the *bushel*, and the *quarter*.

The following are the relations:

Pints.	Quarts.	Gallons.	Pecks.	Bushels.	Quarters.
1	0.5	0.125	0.0625	0.015625	0.001953125
2	1.	0.25	0.125	0.03125	0.00390625
8	4.	1.	0.5	0.125	0.015625
16	8.	2.	1.	0.25	0.03125
64	32.	8.	4.	1.	0.125
512	256.	64.	32.	8.	1.

IV. *Tables of British Weights.*

1.—*Imperial Troy Weight.*

Standard : One cubic inch of distilled water, at 62° Fahrenheit's Thermometer, the Barometer being 30 inches, weighs 252.458 Troy grains.

$$
\begin{array}{llll}
\text{grs.} & \text{dwt.} & & \\
24 = & 1 & & \\
 & & \text{oz.} & \\
480 = & 20 = & 1 & \\
 & & & \text{lb.} \\
5760 = & 240 = & 12 = & 1
\end{array}
$$

Troy weight is used in weighing gold, silver, jewels, &c., and in philosophical experiments.

2.—*Imperial Avoirdupois Weight.*

Standard: The same as in Troy weight, and one avoirdupois pound = 7000 Troy grains.

$$
\begin{array}{llllll}
\text{drs.} & \text{oz.} & & & & \\
16 = & 1 & & & & \\
 & & \text{lb.} & & & \\
256 = & 16 = & 1 & & & \\
 & & & \text{qr.} & & \\
7168 = & 448 = & 28 = & 1 & & \\
 & & & & \text{cwt.} & \\
28672 = & 1792 = & 112 = & 4 = & 1 & \\
 & & & & & \text{ton.} \\
573440 = & 35840 = & 2240 = & 80 = & 20 = & 1
\end{array}
$$

This weight is used for the general purposes of commerce.

V. *Miscellaneous.*

Length.—Gunter's chain = 66 feet = 4 poles = 100 links of 7.92 inches.

 1 fathom = 6 feet; 1 cable length = 120 fathoms.

 1 hand = 4 inches; 1 palm = 3 inches; 1 span = 9 inches.

Solid.—1 cubic yard = 27 cubic feet (B. M.) = 1728 cubic inches.

 1 reduced foot (B. M.) = 1 square foot × 1 inch thick = 144 cubic inches.

 1 perch of masonry = 1 perch (16½ feet) long × 1 foot high × 1½ foot thick = 24.75 cub. feet; 25 cubic feet has generally been adopted for convenience.

 1 cord fire wood = 8 feet long × 4 feet high × 4 feet deep = 128 cubic feet.

 1 chaldron coal = 36 bushels = 57.25 cub. feet.

Paper.—24 sheets = 1 quire.

 20 quires = 1 ream = 480 sheets.

Dimensions of Drawing Paper.

Cap	13 × 16 in.		Elephant	27¾ × 22¼ in.
Demy	19½ × 15½		Columbia	33¾ × 23
Medium	22 × 18		Atlas	33 × 26
Royal	24 × 19		Theorem	34 × 28
Super Royal	27 × 19		Double eleph't	40 × 26
Imperial	29 × 21¼		Antiquarian	52 × 31

Capacities.

A box 16 × 16.8 × 8. in. contains	1 bushel	
12 × 11.2 × 8. "	½ bushel	dry measure
8 × 8.4 × 8. "	1 peck	
6 × 6 × 6.4 "	1 gallon	liquid meas.
4 × 4 × 3.6 "	1 quart	

VI. *French System of Measures.*

The *unit of measures of length* is the *metre.* The unit of *superficial measure* is the *are,* a surface of 10 metres each way, or 100 square metres. The unit of *measures of capacity* is the *litre,* a vessel containing the cube of the tenth part of the metre. The standard temperature is that of melting ice.

The measures of length are:

Myriametre	=	10000 metres.
Kilometre	=	1000
Hectometre	=	100
Metre	=	1
Decimetre	=	0.1
Centimetre	=	0.01
Millimetre	=	0.001

The measures of surface are:

Hectare	=	10000 sq. metres.
Are	=	100
Centiare	=	1

The measures of capacity are:

Kilolitre	=	1000 litres.
Hectolitre	=	100
Decalitre	=	10
Litre	=	1
Decilitre	=	0.1
Centilitre	=	0.01

The unit of solid measure is the *stere* or cube of the metre, equal to 35.31658 English cubic feet.

WEIGHTS AND MEASURES.

Table for converting Metres into Toises and French and English feet and inches.

Met.	Toises.	French.			English.	
		Feet.	In.	Lines.	Feet.	Inches.
1	0.51307	3	0	11.296	3	3.3708
2	1.02615	6	1	10.592	6	6.7416
3	1.53922	9	2	9.888	9	10.1124
4	2.05230	12	3	9.184	13	1.4832
5	2.56537	15	4	8.480	16	4.8539
6	3.07844	18	5	7.776	19	8.2247
7	3.59152	21	6	7.072	22	11.5955
8	4.10459	24	7	6.368	26	2.9663
9	4.61767	27	8	5.664	29	6.3371
10	5.13074	30	9	4.960	32	9.7079
20	10.26148	61	6	9.920	65	7.4158
30	15.39222	92	4	2.880	98	5.1237
40	20.52296	123	1	7.840	131	2.8316
50	25.65370	153	11	0.800	164	0.5395
60	30.78444	184	8	5.760	196	10.2474
70	35.91519	215	5	10.720	229	7.9553
80	41.04593	246	3	3.680	262	5.6632
90	46.17667	277	0	8.640	295	3.3711
100	51.30741	307	10	1.600	328	1.0790
200	102.61481	615	8	3.200	656	2.1580
300	153.92222	923	6	4.800	984	3.2370
400	205.22963	1231	4	6.400	1312	4.3160
500	256.53704	1539	2	8.000	1640	5.3950
600	307.84444	1847	0	9.600	1968	6.4740
700	359.15185	2154	10	11.200	2296	7.5530
800	410.45926	2462	9	0.800	2624	8.6320
900	461.76667	2770	7	2.400	2952	9.7110
1000	513.07407	3078	5	4.000	3280	10.7900
2000	1026.14815	6156	10	8.000	6561	9.5800
3000	1539.22222	9235	4	0.000	9842	8.3700
4000	2052.29630	12313	9	4.000	13123	7.1600
5000	2565.37037	15392	2	8.000	16404	5.9500
6000	3078.44444	18470	8	0.000	19685	4.7400
7000	3591.51852	21549	1	4.000	22966	3.5300
8000	4104.59259	24627	6	8.000	26247	2.3200
9000	4617.66667	27706	0	0.000	29528	1.1100
10000	5130.74074	30784	5	4.000	32808	11.9000

Log. to reduce metres to Eng. feet = 0.5159929.

Table for converting English Feet into French Toises, Metres, and Feet.

English feet.	Toises.	Metres.	French. Feet.	In.	Lines.
1	0.15638	0.30479	0	11	3.114
2	0.31276	0.60959	1	10	6.228
3	0.46915	0.91438	2	9	9.343
4	0.62553	1.21918	3	9	0.457
5	0.78191	1.52397	4	8	3.571
6	0.93829	1.82877	5	7	6.685
7	1.09468	2.13356	6	6	9.799
8	1.25106	2.43836	7	6	0.913
9	1.40744	2.74315	8	5	4.028
10	1.56382	3.04794	9	4	7.142
20	3.12764	6.09589	18	9	2.284
30	4.69146	9.14383	28	1	9.425
40	6.25529	12.19178	37	6	4.567
50	7.81911	15.23972	46	10	11.709
60	9.38293	18.28767	56	3	6.851
70	10.94675	21.33561	65	8	1.993
80	12.51057	24.38536	75	0	9.134
90	14.07439	27.43150	84	5	4.276
100	15.63822	30.47945	93	9	11.418
200	31.27643	60.95850	187	7	10.836
300	46.91465	91.43835	281	5	10.254
400	62.55286	121.91780	375	3	9.672
500	78.19108	152.39725	469	1	9.090
600	93.82929	182.87670	562	11	8.508
700	109.46751	213.35615	656	9	7.926
800	125.10572	243.83559	750	7	7.344
900	140.74394	274.31504	844	5	6.762
1000	156.38215	304.79449	938	3	6.180
2000	312.76431	609.58899	1876	7	0.360
3000	469.14646	914.38348	2814	10	6.539
4000	625.52861	1219.17797	3753	2	0.719
5000	781.91076	1523.97246	4691	5	6.899
6000	938.29292	1828.76696	5629	9	1.079
7000	1094.67507	2133.56145	6568	0	7.259
8000	1251.05722	2438.35594	7506	4	1.438
9000	1407.43937	2743.15044	8444	7	7.618
10000	1563.82153	3047.94493	9382	11	1.798

Log. to reduce English feet to metres = 9.4840071.

Table of relations between the Linear Measures of several countries, with corresponding logarithms.

VII. *Foreign Measures of Length.*

Metre.	France. Paris foot.	England. Russia. foot.	Prussia. Denmark. foot.	Bavaria. foot.	Saxony. foot.	Baden. Switzerland. foot.	Austria. Vienna foot.	Spain. Mexico. foot.
1	3.078444 0.4883313	3.280899 0.5159929	3.186199 0.5032730	3.426310 0.5348266	3.531197 0.5479220	3.333333 0.5228787	3.163446 0.5001605	3.537877 0.5487427
0.3248394 9.5116687	1	1.065765 0.0276616	1.035003 0.0149417	1.113000 0.0464954	1.147072 0.0595907	1.082798 0.0345475	1.027612 0.0118292	1.149242 0.0604114
0.3047945 9.4840071	0.938293 9.9723384	1	0.971136 9.9872801	1.044320 0.0188337	1.076290 0.0319291	1.015982 0.0068859	0.964201 9.9841676	1.078325 0.0327498
0.3138535 9.4967270	0.966181 9.9850583	1.029722 0.0127199	1	1.075359 0.0315536	1.108279 0.0446490	1.046178 0.0196058	0.922859 9.9968875	1.110375 0.0454697
0.2918592 9.4651734	0.898472 9.9535047	0.957561 9.9811663	0.929922 9.9684464	1	1.030612 0.0130954	0.972864 9.9880521	0.923281 9.9653339	1.032562 0.0139161
0.2831901 9.4520780	0.871785 9.9404093	0.929118 9.9680709	0.902300 9.9553510	0.970297 9.9869046	1	0.943967 9.9749567	0.895856 9.9522385	1.001892 0.0008207
0.3000000 9.4771213	0.923533 9.9654525	0.984270 9.9931141	0.955860 9.9803942	1.027893 0.0119479	1.059359 0.0250433	1	0.949034 9.9772817	1.061361 0.0258630
0.3161109 9.4998395	0.973130 9.9881708	1.037128 0.0158324	1.007193 0.0031125	1.083094 0.0346661	1.116250 0.0477615	1.053703 0.0227183	1	1.118361 0.0485822
0.2826553 9.4512573	0.870139 9.9395886	0.927364 9.9672502	0.900597 9.9545303	0.968465 9.9860839	0.998112 9.9991793	0.942184 9.9741360	0.894165 9.9514178	1

VIII. *Table of relations between Itinerary Measures of several countries, with the corresponding logarithms.*

	France. Myriametre =10000 M.	England. Stat. mile =5280'.	Prussia. Denmark mile =24000'.	Austria mile =24000'.	Russia verst =3500'.	Spain. Mexico Jud. league =15000'.	Germany. Geo. mile 15=1 deg.	England. France. Naut. l'gue 20=1 deg.
England. France. Naut. l'gue 20=1 deg.	1.796907 / 0.2545256	0.289179 / 9.4611666	1.353518 / 0.1314639	1.363253 / 0.1345764	0.191691 / 9.2826008	0.761868 / 9.8818742	1.333333 / 0.1249387	1
Germany. Geo. mile 15=1 deg.	1.347680 / 0.1295869	0.216884 / 9.3362279	1.015138 / 0.0065251	1.022440 / 0.0096376	0.143768 / 9.1576620	0.571394 / 9.7569355	1	0.750000 / 9.8750613
Spain. Mexico Jud. league =15000'.	2.358584 / 0.3726514	0.379570 / 9.5792924	1.776600 / 0.2495897	1.789379 / 0.2527022	2.516092 / 0.4007266	1	1.750107 / 0.2430645	1.312580 / 0.1181258
Russia verst =3500'.	9.373997 / 0.9719248	1.508571 / 0.1785659	7.060950 / 0.8488631	7.111736 / 0.8519756	1	0.397442 / 9.5992734	6.955654 / 0.8423380	5.216740 / 0.7173992
Austria mile =24000'.	1.318103 / 0.1199492	0.212124 / 9.3265903	0.992859 / 9.9968875	1	0.140613 / 9.1480244	0.558853 / 9.7472978	0.978053 / 9.9903624	0.733540 / 9.8654236
Prussia. Denmark mile =24000'.	1.327583 / 0.1230617	0.213650 / 9.3297028	1	1.007193 / 0.0031125	0.141624 / 9.1511369	0.562873 / 9.7504103	0.985088 / 9.9934749	0.738816 / 9.8685361
England. Stat. mile =5280'.	6.213824 / 0.7933590	1	4.680554 / 0.6702972	4.714219 / 0.6734097	0.662879 / 9.8214341	2.634556 / 0.4207076	4.610755 / 0.6637721	3.458067 / 0.5388334
France. Myriametre =10000 M.	1	0.1609315 / 9.2066410	0.7532485 / 9.8769383	0.7586663 / 9.8800508	0.1066781 / 9.0280752	0.4239831 / 9.6273486	0.7420158 / 9.8704131	0.5565118 / 9.7454744

1 English or French geographical mile = 1-60 of a degree of longitude at the Equator = 2028.7 English yards.

	Eng. stat. miles.			Eng. stat. miles.
Modern Roman mile	= 0.925	Portugal league	=	3.841
Tuscan mile	= 1.027	Flanders league	=	3.900
Old Scottish mile	= 1.127	Spanish common league	=	4.214
Irish mile	= 1.273	Hungarian mile	=	5.178
French posting league	= 2.422	Swedish mile	=	6.648

Comparison of French and English Measures.

Metre	39.37079 inches.
"	3.28089 feet.
"	1.09363 yards.
Kilometre	0.62138 miles.
Myriametre . . .	6.2138 miles.

Square metre . . .	1.196033 square yards.
Are	.119.6033 square yards.
Hectare	2.471143 acres.

Litre	1.760773 pints.
"	0.220096 gallons.
Decalitre	2.200967 gallons.
Hectolitre	22.009668 gallons.

Gramme	15.438 grains, troy.
"	0.032 ounce, troy.
Kilogramme . . .	2.680 pounds, troy.
" . . .	2.205 pounds, avoirdupois.

IX. *Spanish and Mexican Measures of length.*

1 Castilian foot	= 11.1284	English inches.
3 Castilian feet = 1 vara	= 33.3852	English inches.
	= 0.927365	English yards.
5000 varas = 1 judicial league	= 4637.	English yards.

X. *Specific Gravities.*

Substance.	Specific gravity.	Weight of 1 cub. inch.	Substance.	Specific gravity.	Weight of 1 cub. in.
		Lbs.			Lbs.
Brass, (cast).......	8.396	0.3037	Sand............	1.800	0.0652
Bronze, (gun metal)	8.700	0.3147	Stone, (common)	2.520	0.0911
Copper, (cast)......	8.788	0.3179	Wood, ash......	0.722	0.0261
Iron, (bar).........	7.788	0.2817	" cypress ..	0.441	0.0160
Iron, (cast)........	7.207	0.2607	" hickory...	0.838	0.0303
Lead, (cast)........	11.352	0.4106	" oak.......	0.687	0.0248
Tin, (cast).........	7.291	0.2637	" pine	0.541	0.0196
Bricks............	1.900	0.0690	Coal, (bitumin's)	1.270	0.0460
Earth, (common)...	1.500	0.0543	Water, (distilled)	1.000	0.0361

The weight of dry atmospheric air at the temperature of 32°, the barometer being at 30 in., is $\frac{1}{770}$ of that of distilled water.

The weight of a cubic foot of distilled water *at the maximum density* being nearly 1000 ounces avoirdupois, the specific gravity of a solid or liquid body expresses the weight of a cubic foot, in ounces; therefore the weight of such a body in ounces will be found by multiplying its contents in cubic feet by its specific gravity.

According to Mr. Hassler's comparisons, the weight of a cubic foot of water at its maximum density, the barometer being at 30 in., is 998.068 oz.

According to the British imperial standards, the weight of a cubic foot of water, at 62°, the barometer being at 30 in., is 997.136 oz.; this would give for the cubic foot of water, at the maximum density, 998.224 oz.

By the investigations of Prof. R. S. McCulloch, the maximum density of water is at the temperature of 39°.6 Fahr.; this agrees very nearly with Mr. Hassler's determination of the maximum density, 39°.83.

XI. *Analytical Expressions for different Lines, Surfaces, and Solids.*

1. *Lines.*

Circle.   Ratio of circumference to diameter $= 3.1415926$

$= \frac{355}{113}$ nearly.

Length of an arc $= \frac{a\,\pi\,r}{180}$; r being the radius of the circle, and a the number of degrees in the arc ; or, nearly $= \frac{8\,c' - c}{3}$; c being the chord of the arc, and c' (the chord of half the arc) $= \sqrt{\frac{1}{4}\,c^2 + \text{versine}^2}$.

Ellipse.   Circumference $= \frac{199}{200}\,\pi\,\sqrt{\frac{1}{2}\,(a^2 + b^2)}$ nearly; a and b being the axes.

Parabola. Length of an arc, commencing at vertex $= \sqrt{\frac{4\,a^2}{3} + \sqrt{b}}$ nearly; a being the abcissa, and b the ordinate.

2. *Surfaces.*

1. Triangle in terms of—

its base and its altitude . . . . $= \dfrac{b\,A}{2}$

two sides and the included angle . . . . . $= \dfrac{a\,b\,\sin C}{2}$

its three sides . . $= [\,s\,(s-a)\,(s-b)\,(s-c)\,]^{\frac{1}{2}}$

where A $=$ the altitude; a, b, $c =$ the three sides, and C the angle included between a and b ; $s = \dfrac{a+b+c}{2}$

2. Parallelogram in terms of—

its base and its altitude . . . . $= b\,A$

two sides and the included angle . . . . . $= a\,b\,\sin C$

two sides and their corresponding diagonal

$$= 2\,[\,s\,(s-a)\,(s-b)\,(s-c)\,]^{\frac{1}{2}}$$

where C $=$ the angle included between two adjacent sides a, b;

$c =$ the diagonal opposite, and $s = \dfrac{a+b+c}{2}$.

3. Trapezium in terms of—

its two parallel bases and its altitude . . . $= \dfrac{B+b}{2}\,A$

its two parallel bases, one of its oblique sides and the angle between one of these bases and this side . . . . . $\Big\} = \dfrac{B+b}{2}\,l\,\sin C$

where A $=$ the distance between the two parallel bases B, b; $l =$ the length of one of the oblique sides, and C the angle between one of these bases and this side.

4. Any Quadrilateral $=$ half the product of its two diagonals multiplied by the sine of the included angle.

5. Regular Polygon . . . . $= \dfrac{n\left(\frac{a}{2}\right)^2}{\tan \dfrac{180°}{n}}$

where $n =$ the number of sides ; $a =$ the length of one of them.

6. Circle . . . . $= \pi\,R^2$

7. Ellipse. . . . . $= \pi\,a\,b$
   a and b being the semi-axes.

8. Right cylinder, exclusive of its bases . $= 2\,\pi\,R\,A$

9. Sphere. . . . . $= 4\,\pi\,R^2$

10. Zone . $= 4\,\pi\,R^2 \sin \tfrac{1}{2}\,(\,L' - L\,)\,\cos \tfrac{1}{2}\,(\,L' + L\,)$

11. Spherical Quadrilateral, formed by two parallels of
Latitude and two meridians

$$\stackrel{\cdot}{=} \frac{\pi}{90^\circ} \, (\, M' - M \,) \, R^2 \sin \tfrac{1}{2} \, (\, L' - L \,) \cos \tfrac{1}{2} \, (\, L' + L \,)$$

where R $=$ the radius of the sphere; L, L$'$ $=$ the latitudes
of the bases of the zone, $+$ when North, $-$ South; M$'$, M
$=$ the longitudes of the extreme meridians of the quadrilateral.   (M$'$ $-$ M) being expressed in degrees and decimals.

In the place of R, the normal N, of the mean Latitude
$\left(\dfrac{L' + L}{2} \right)$, can be used.

12. Right cone . . . . $= \pi \, R \, L$

13. Frustrum of cone with parallel bases $= \pi \, l \, (\, R + r \,)$

When R and $r =$ the radii of the bases of these solids,
L and $l =$ the lengths of their generating elements.

3. *Solids.*

14. Prism . . . . $= B \, A$
     where B $=$ the area of the base, A $=$ the altitude.

15. Rectangular parallelopiped . . . $= p \times q \times r$
    Cube . . . . $= p^3$
    where p, q, r, $=$ the lengths of the three contiguous
    edges.

16. Pyramid . . . . $= \dfrac{B \, A}{3}$

    The area B being found from No. 5.

17. Right cylinder . . . . $= \pi \, R^2 \, A$

18. Right cone . . . . $= \tfrac{1}{3} \pi \, R^2 \, A$

19. Sphere . . . . $= \tfrac{4}{3} \pi \, R^3$

20. Prismoid, or solid figure, similar to that which is formed in excavations or embankments of roads; terminated by parallel cross sections. Solid content = area of each end, added to four times the middle area, and the sum multiplied by the length divided by 6, or

$$= \left\{ (b + r\,h')\,h' + (b + r\,h)\,h + 4\left(b + r = \frac{h + h'}{2}\right)\frac{h + h'}{2} \right\}\frac{l}{6}$$

where b = the breadth at the bottom of the cutting

h = the perpendicular depth of cutting at higher end

h' = the perpendicular depth of cutting at lower end

l = the length of the solid

r = the ratio of the perpendicular height of the slope to its horizontal base.

Lengths of Circular Arcs,
Taking the base of Segments as unity.

Ver. Sin.	Length.	Ver. Sin.	Length.	Ver. Sin.	Length.	Ver. Sin.	Length.	Ver. Sin.	Length.
.01	1.000	.11	1.032	.21	1.114	.31	1.239	.41	1.401
.02	"	.12	1.038	.22	1.124	.32	1.254	.42	1.418
.03	"	.13	1.044	.23	1.135	.33	1.269	.43	1.437
.04	"	.14	1.051	.24	1.147	.34	1.284	.44	1.455
.05	"	.15	1.059	.25	1.159	.35	1.300	.45	1.474
.06	1.006	.16	1.067	.26	1.171	.36	1.316	.46	1.493
.07	1.014	.17	1.075	.27	1.184	.37	1.332	.47	1.512
.08	1.018	.18	1.084	.28	1.197	.38	1.349	.48	1.531
.09	1.020	.19	1.093	.29	1.212	.39	1.366	.49	1.551
.10	1.026	.20	1.103	.30	1.225	.40	1.383	.50	1.571

XII. *Hydrometry.*

1. To determine the mean velocity of a stream from observations of the velocity at its surface.

Let α = the observed surface velocity, in inches,

β = the bottom velocity,

γ = the mean velocity,

$$\beta = (\sqrt{\alpha} - 1)^2, \qquad\qquad \gamma = \frac{\alpha + \beta}{2}$$

$$\gamma = \frac{\alpha + (\sqrt{\alpha} - 1)^2}{2}\ .$$

Prony has given the very simple formula

$$\gamma = 0.816458\ \alpha$$

which is, perhaps, more correct than the above of Dubuat.

2. In open streams which are flowing with *an uniform motion,* calling

ω, the area of the section of the stream,

$\varkappa$, that portion of the perimeter of the section of the bed, which is in contact with the water,

I, the fall divided by the length,

ν, the mean velocity per second $= \dfrac{Q}{\omega}$,

Q, the discharge per second,

R, the *hydraulic depth,* or $\dfrac{\omega}{\varkappa}$,

(the unit being English feet), the relations, according to Eytelwein, between these several quantities, may be expressed by the following,

$$0.0000242651\,\nu + 0.0001114155\,\nu^2 = R\,I.$$

Whence

$$\nu = -0.1088941604 + \sqrt{0.0118580490 + 8975.414285\,R\,I.}$$

And $0.0000242651\,Q\,\omega + 0.0001114155\,Q^2 = \dfrac{\omega^3}{\varkappa}\,I.$

Whence

$$Q= \sqrt{\left(8975.414285 \frac{\omega^3}{x} I + 0.01213425\, \omega^2\right)} - 0.1088942\, \omega$$

Log 0.01213425   = 8.0840130
Log 0.1088942   = 9.0370065
Log 8975.414285   = 3.9530545
Log 0.0000242651 = 5.3849821
Log 0.0001114155 = 6.0469456

To realize in practice what would be called uniform motion, the canal or stream should be straight, and with the same section and inclination from one end to the other. In proportion as it varies from these conditions, we may expect to find the formula in fault.

Table of Surface, Bottom and Mean Velocities.

VELOCITY IN INCHES.					
Surface.	Bottom.	Mean.	Surface.	Bottom.	Mean.
5	1.527	3.263	55	41.167	48.083
10	4.675	7.337	60	45.508	52.754
15	8.254	11.627	65	49.875	57.436
20	12.055	16.027	70	54.266	62.133
25	16.000	20.500	75	58.679	66.839
30	20.045	25.022	80	63.111	71.555
35	24.167	29.583	85	67.561	76.280
40	28.350	34.175	90	72.026	81.006
45	32.583	38.791	95	76.506	85.753
50	36.857	43.428	100	81.000	90.500

XIII. *To trace Railroad Curves by means of deflections.*

General Propositions.

1. The angle formed by a tangent and a chord is equal to half the angle at the centre of the circle, subtended by the chord.

2. The angle of deflection formed by any two equal chords meeting at the circumference, is equal to the angle at the centre, subtended by either chord.

3. A line bisecting the angle of deflection formed by any two equal chords, is a tangent to the arc at the point where the two chords meet.

4. If an arc of a circle be subdivided into any number of equal parts, and lines be drawn from the several points of subdivision so as to meet at any point in the circumference, these several lines will form equal angles at the point of meeting, and the angles thus formed will be respectively measured by one half the subdivided arc.

TABLE 1.

Table of deflections for chords and tangents with radii and versed sines corresponding.

Angle of deflection between tangent and chord.	Angle of deflection for chords of 100 feet.	Corresponding radius of circle.	Length of versed sine for chords of 100 feet.	Angle of deflection between tangent and chord.	Angle of deflection for chords of 100 feet	Corresponding radius of circle.	Length of versed sine for chords of 100 feet.
D. M.	D. M.	feet.		D. M.	D. M.	feet.	
° ′	° ′			° ′	° ′		
0.15	0.30	11460.	.106	3.45	7.30	764.	1.635
0.30	1.00	5730.	.217	4.00	8.00	716.2	1.744
0.45	1.30	3820.	.328	4.15	8.30	674.1	1.853
1.00	2.00	2865.	.435	4.30	9.00	636.6	1.960
1.15	2.30	2292.	.545	4.45	9.30	603.1	2.070
1.30	3.00	1910.	.655	5.00	10.00	573.	2.180
1.45	3.30	1637.1	.762	5.15	10.30	545.7	2.286
2.00	4.00	1432.5	.872	5.30	11.00	520.9	2.394
2.15	4.30	1273.3	.981	5.45	11.30	498.2	2.505
2.30	5.00	1146.	1.090	6.00	12.00	477.5	2.613
2.45	5.30	1041.8	1.199	6.15	12.30	458.4	2.722
3.00	6.00	955.	1.309	6.30	13.00	440.7	2.828
3.15	6.30	881.5	1.416	6.45	13.30	424.4	2.940
3.30	7.00	818.5	1.525	7.00	14.00	409.2	3.048
3.45	7.30	764.	1.635	7.15	14.30	395.2	3.157

TABLE 2.

Table showing, for arcs of different radii, the lengths of lines of deflection from a tangential point to points on the arc 100 feet apart, with the angles of deflection and versed sines corresponding.

Angle of deflection from Tangent.	Length of line of deflection.	Versed sine for line of deflection.	Angle of deflection from Tangent.	Length of line of deflection.	Versed sine for line of deflection.
deg. min.	feet.	feet.	deg. min.	feet.	feet.
2° Radius 2865 ft.			2½° Rad. 2292 ft.		
1°.00′	100.00	.43	1°.15′	100.00	.54
2 .00	199.97	1.74	2 .30	199.95	2.18
3 .00	299.88	3.93	3 .45	299.81	4.90
4 .00	399.70	6.98	5 .00	399.53	8.72
5 .00	499.39	10.90	6 .15	499.05	13.62
3° Rad. 1910 ft.			3½° Rad. 1637.1 ft.		
1°.30′	1C0.00	.65	1°.45′	100.00	.76
3 .00	199.93	2.62	3 .30	199.91	3.05
4 .30	299.73	5.89	5 .15	299.63	6.87
6 .00	399.32	10.46	7 .00	399.07	12.20
7 .30	498.63	16.34	8 .45	498.14	19.05
4° Rad. 1432.5 ft.			4½° Rad. 1273.3 ft.		
2°.00′	100.00	.87	2°.15′	100.00	.98
4 .00	199.88	3.49	4 .30	199.85	3.92
6 .00	299.51	7.84	6 .45	299.38	8.90
8 .00	398.78	13.93	9 .00	398.46	15.67
10 .00	497.57	21.75	11 .15	496.92	24.46
5° Rad. 1146 ft.			5½° Rad. 1041.8 ft.		
2°.30′	100.00	1.09	2°.45′	100.00	1.20
5 .00	199.81	4.36	5 .30	199.77	4.79
7 .30	299.24	9.80	8 .15	299.08	10.77
10 .00	398.10	17.41	11 .00	397.70	19.12
12 .30	496.20	27.16	13 .45	495.41	29.83

Table 2—Continued.

Angle of deflection from Tangent.	Length of line of deflection.	Versed sine for line of deflection.	Angle of deflection from Tangent.	Length of line of deflection.	Versed sine for line of deflection.
deg. min.	feet.	feet.	deg. min.	feet.	feet.
6° Rad. 955 ft.			6½° Rad. 881 ft.		
3°.00′	100.00	1.31	3°.15′	100.00	1.41
6 .00	199.73	5.23	6 .30	199.68	5.66
9 .00	298.90	11.75	9 .45	298.71	12.72
12 .00	397.26	20.86	13 .00	396.79	22.58
15 .00	494.53	32.54	16 .15	493.59	35.19
7° Rad. 818.5 ft.			7½° Rad. 764 ft.		
3°.30′	100.00	1.52	3°.45′	100.00	1.63
7 .00	199.63	6.09	7 .30	199.57	6.53
10 .30	298.51	13.69	11 .15	298.29	14.68
14 .00	396.28	24.30	15 .00	395.73	26.03
17 .30	492.57	37.86	18 .45	491.47	40.54
8° Rad. 716.2 ft.			8½° Rad. 674.1 ft.		
4°.00′	100.00	1.74	4°.15′	100.00	1.85
8 .00	199.51	6.97	8 .30	199.48	7.40
12 .00	298.05	15.64	12 .45	297.84	16.62
16 .00	395.14	27.73	17 .00	394.57	29.45
20 .00	490.28	43.18	21 .15	489.13	45.83
9° Rad. 636.6 ft.			9½° Rad. 603.1 ft.		
4°.30′	100.00	1.96	4°.45′	100.00	2.07
9 .00	199.39	7.83	9 .30	199.31	8.27
13 .30	297.54	17.57	14 .15	297.26	18.55
18 .00	393.86	31.13	19 .00	393.16	32.85
22 .30	487.75	48.41	23 .45	486.36	51.07
10° Rad. 573 ft.			10½° Rad. 545.7 ft.		
5°.00′	100.00	2.18	5°.15′	100.00	2.28
10 .00	199.24	8.70	10 .30	199.16	9.12
15 .00	296.96	19.52	15 .45	296.65	20.46
20 .00	392.42	34.55	21 .00	391.65	36.19
25 .00	484.90	53.68	26 .15	483.37	56.20

Table 2—Continued.

Angle of deflection from Tangent.	Length of line of deflection.	Versed sine for line of deflection.	Angle of deflection from Tangent.	Length of line of deflection.	Versed sine for line of deflection.
deg. min.	feet.	feet.	deg. min.	feet.	feet.
11° Rad. 520.9 ft.			11½° Rod. 498.2 ft.		
5°.30′	100.00	2.39	5°.45′	100.00	2.50
11 .00	199.08	9.56	11 .30	198.99	9.99
16 .30	296.33	21.41	17 .15	295.99	22.40
22 .00	390.84	37.86	23 .00	390.00	39.58
27 .30	481.76	58.75	28 .45	480.10	61.39
12° Rad. 477.5 ft.			12½° Rad. 458.4 ft.		
6°.00′	100.00	2.61	6°.15′	100.00	2.72
12 .00	198.90	10.42	12 .30	198.81	10.85
18 .00	295.63	23.34	18 .45	295.26	24.30
24 .00	389.12	41.24	25 .00	388.20	42.91
30 .00	478.34	63.90	31 .15	476.52	66.45
13° Rad. 440.7 ft.			13½°		
6°.30′	100.00	2.82	6°.45′	100.00	2.94
13 .00	198.71	11.27	13 .30	198.61	11.71
19 .30	294.87	25.23	20 .15	294.47	26.20
26 .00	387.24	44.53	27 .00	386.25	46.21
32 .30	474.63	68.90	33 .45	472.68	71.45
14°			14½°		
7°.00′	100.00	3.04	7°.15′	100.00	3.15
14 .00	198.51	12.14	14 .30	198.40	12.58
21 .00	294.06	27.16	21 .45	293.63	28.12
28 .00	385.23	47.87	29 .00	384.16	49.52
35 .00	470.65	73.96	36 .15	468.55	76.45
42 .00	549.06	105.05	43 .30	545.45	108.47
49 .00	619.28	140.67	50 .45	613.63	145.08
56 .00	680.27	180.29	58 .00	671.99	185.65
63 .00	731.12	223.32	65 .15	719.61	229.63
70 .00	771.07	269.11	72 .30	755.73	276.22

TABLE 3.

Table of ordinates to circular arcs on a chord of 100 feet.

Angle of deflec-tion for chords.	ABSCISSA IN FEET.									
	5. 95.	10. 90.	15. 85.	20. 80.	25. 75.	30. 70.	35. 65.	40. 60.	45. 55.	50.
1.00	.04	.07	.11	.14	.16	.18	.20	.21	.21	.22
1.30	.06	.10	.16	.21	.25	.28	.30	.31	.32	.33
2.00	.08	.14	.22	.28	.33	.37	.40	.41	.43	.44
2.30	.11	.20	.28	.35	.41	.46	.50	.52	.54	.55
3.00	.13	.24	.33	.42	.49	.55	.59	.63	.65	.66
3.30	.15	.28	.39	.49	.57	.64	.69	.74	.76	.77
4.00	.17	.32	.44	.56	.66	.73	.79	.84	.86	.87
4.30	.19	.36	.50	.63	.74	.83	.89	.95	.97	.98
5.00	.21	.40	.56	.70	.82	.92	.99	1.05	1.08	1.09
5.30	.23	.43	.61	.77	.91	1.01	1.09	1.16	1.19	1.20
6.00	.25	.47	.67	.84	.99	1.10	1.19	1.26	1.30	1.31
6.30	.27	.51	.72	.91	1.07	1.19	1.29	1.37	1.41	1.42
7.00	.29	.55	.78	.98	1.15	1.28	1.39	1.47	1.52	1.53
7.30	.31	.59	.83	1.05	1.23	1.38	1.49	1.57	1.63	1.64
8.00	.33	.63	.89	1.13	1.31	1.47	1.59	1.68	1.74	1.75
8.30	.35	.67	.94	1.20	1.39	1.56	1.69	1.78	1.84	1.86
9.00	.37	.71	1.00	1.26	1.47	1.65	1.78	1.89	1.95	1.97
9.30	.40	.75	1.06	1.33	1.56	1.74	1.88	1.99	2.05	2.08
10.00	.42	.79	1.11	1.40	1.64	1.83	1.98	2.10	2.15	2.19
10.30	.44	.82	1.17	1.47	1.72	1.93	2.08	2.20	2.26	2.29
11.00	.46	.86	1.22	1.54	1.80	2.02	2.18	2.31	2.36	2.40
11.30	.48	.90	1.28	1.61	1.88	2.11	2.28	2.41	2.47	2.51
12.00	.50	.94	1.34	1.68	1.97	2.20	2.38	2.52	2.57	2.62
12.30	.52	.98	1.39	1.75	2.05	2.29	2.48	2.62	2.68	2.72
13.00	.54	1.02	1.45	1.82	2.13	2.38	2.58	2.73	2.78	2.84
13.30	.56	1.06	1.50	1.89	2.21	2.48	2.68	2.83	2.89	2.94
14.00	.58	1.10	1.56	1.96	2.29	2.57	2.78	2.93	3.00	3.05
14.30	.60	1.12	1.61	2.02	2.37	2.65	2.87	3.03	3.13	3.16

XIV. *Land Surveying with Compass and Chain.*

To calculate the Area or Content of Land.

If the sum of each adjacent pair of distances perpendicular to a meridian (*departures*) assumed without the survey, be multiplied by the northing or southing between them, in succession round the figure in the same order, the difference between the sum of the *north* products and the sum of the *south* products will be double the area of the tract.

The *meridian distance* of a course is the distance of the middle point of that course from an assumed meridian.

Hence—The double meridian distance of the first course is equal to its departure.

And the double meridian distance of any course is equal to the double meridian distance of the preceding course, plus its departure, plus the departure of the course itself, having regard to the algebraic sign of each.

Then to find the area—

1. Multiply the double meridian distance of each course by its northing or southing.

2. Place all the *plus* products in one column, and all the *minus* products in another.

3. Add up each column separately and take their difference. This difference will be *double* the area of the land.

In *balancing* the work, the error for each particular course is found by the proportion—

As the sum of the courses, is to the error of latitude, (or departure,) so is each particular course, to its correction.

When a bearing is due east or west, the error of latitude is nothing, and the course must be subtracted from the sum of the courses before balancing the columns of latitude. And so with the departures.

EXAMPLE.—It is required to find the content of a piece of land, of which the following are the field notes :

Sta.	Course.	Dist.		Sta.	Course.	Dist.
1.	N. 46½° W. 20.	chains.		4.	S. 56 ° E. 27.60	chains.
2.	N. 51¾° E. 13.80	"		5.	S. 33½° W. 18.80	"
3.	East 21.25	"		6.	N. 74½° W. 30.95	"

Calculation.

Stations.	Courses.	Dist. chains	DIFF. LAT. N. +	DIFF. LAT. S. −	DEPARTURE. E. +	DEPARTURE. W. −	BALANCED. Lat.	BALANCED. Dep.	D.M.D. +	Area +	Area −
1	N. 46½ W.	20.00	13.77	-	-	14.51	+ 13.88	— 14.56	14.56	202.0928	
2	N. 51¾° E.	13.80	8.54	-	10.84	-	+ 8.61	+ 10.81	10.81	93.0741	
3	East	21.25	-	-	21.25	-	-	+ 21.20	42.82	-	
4	S. 56° E.	27.60	-	15.44	22.88	-	— 15.29	+ 22.82	86.84	-	1327.7836
5	S. 33½° W.	18.80	-	15.72	-	10.31	— 15.63	— 10.36	99.30	-	1552.0590
6	N. 74½° W.	30.95	8.27	-	-	29.83	+ 8.43	— 29.91	59.03	497.6229	
Sums . . .		132.40	30.58	31.16	54.97	54.65				792.7898	2879.8426
				30.58		54.65					792.7898

Error in Northing . . . 0 58 | 0.32 Error in Westing . . . . 2)2087.0528

Answer 104 A. 1 R. 16 P. 1043.5264

100.000 square links of Gunter's chain = 1 acre.

XIII. *Traverse Table,*
Showing differences of Latitude and the Departures.

Minutes.	Distance.	0° Lat.	0° Dep.	1° Lat.	1° Dep.	2° Lat.	2° Dep.	Distance.	Minutes.
	1	1.00000	0.00000	0.99984	0.01745	0.99939	0.03490	1	
	2	2.00000	0.00000	1.99969	0.03490	1.99878	0.06980	2	
	3	3.00000	0.00000	2.99954	0.05235	2.99817	0.10470	3	
	4	4.00000	0.00000	3.99939	0.06980	3.99756	0.13960	4	
0′	5	5.00000	0.00000	4.99923	0.08726	4.99695	0.17450	5	60′
	6	6.00000	0.00000	5.99908	0.10471	5.99634	0.20940	6	
	7	7.00000	0.00000	6.99893	0.12216	6.99573	0.24430	7	
	8	8.00000	0.00000	7.99878	0.13961	7.99512	0.27920	8	
	9	9.00000	0.00000	8.99862	0.15707	8.99451	0.31410	9	
	1	0.99999	0.00436	0.99976	0.02181	0.99922	0.03925	1	
	2	1.99998	0.00872	1.99952	0.04363	1.99845	0.07851	2	
	3	2.99997	0.01308	2.99928	0.06544	2.99768	0.11777	3	
15′	4	3.99996	0.01745	3.99904	0.08725	3.99691	0.15703	4	45′
	5	4.99995	0.02181	4.99881	0.10907	4.99614	0.19629	5	
	6	5.99994	0.02617	5.99857	0.13089	5.99537	0.23555	6	
	7	6.99993	0.03054	6.99833	0.15270	6.99460	0.27481	7	
	8	7.99992	0.03490	7.99809	0.17452	7.99383	0.31407	8	
	9	8.99991	0.03926	8.99785	0.19633	8.99306	0.35333	9	
	1	0.99996	0.00872	0.99965	0.02617	0.99904	0.04361	1	
	2	1.99992	0.01745	1.99931	0.05235	1.99809	0.08723	2	
	3	2.99988	0.02617	2.99897	0.07853	2.99714	0.13085	3	
	4	3.39984	0.03490	3.99862	0.10470	3.96619	0.17447	4	
30′	5	4.99981	0.04363	4.99828	0.13088	4.99524	0.21809	5	30′
	6	5.99977	0.05235	5.99794	0.15706	5.99428	0.26171	6	
	7	6.99973	0.06108	6.99760	0.18323	6.99333	0.30533	7	
	8	7.99969	0.06981	7.99725	0.20941	7.99238	0.34895	8	
	9	8.99965	0.07853	8.99691	0.23559	8.99143	0.39257	9	
	1	0.99991	0.01308	0.99953	0.03053	0.99884	0.04797	1	
	2	1.99982	0.02617	1.99906	0.06107	1.99769	0.09595	2	
	3	2.99974	0.03926	2.99860	0.09161	2.99654	0.14393	3	
	4	3.39965	0.05235	3.99813	0.12215	3.99539	0.19191	4	
45′	5	4.99957	0.06544	4.99766	0.15269	4.99424	0.23989	5	15′
	6	5.99948	0.07853	5.99720	0.18323	5.99309	0.28786	6	
	7	6.99940	0.09162	6.99673	0.21376	6.99193	0.33584	7	
	8	7.99931	0.10471	7.99626	0.24430	7.99078	0.38382	8	
	9	8.99922	0.11780	8.99580	0.27484	8.98963	0.43180	9	
Minutes.	Distance.	Dep.	Lat.	Dep.	Lat.	Dep.	Lat.	Distance.	Minutes.
		89°		88°		87°			

5

Differences of Latitude and Departures—Continued.

Minutes	Distance	3° Lat.	3° Dep.	4° Lat.	4° Dep.	5° Lat.	5° Dep.	Distance	Minutes
	1	0.99863	0.05233	0.99756	0.06975	0.99619	0.08715	1	
	2	1.99726	0.10467	1.99512	0.13951	1.99238	0.17431	2	
	3	2.99589	0.15700	2.99269	0.20926	2.98858	0.26146	3	
	4	3.99452	0.20934	3.99025	0.27902	3.98477	0.34862	4	
0'	5	4.99315	0.26168	4.98782	0.34878	4.98097	0.43577	5	60'
	6	5.99178	0.31401	5.98538	0.41853	5.97716	0.52293	6	
	7	6.99041	0.36635	6.98294	0.48829	6.97336	0.61008	7	
	8	7.98904	0.41868	7.98051	0.55805	7.96955	0.69724	8	
	9	8.98767	0.47102	8.97807	0.62780	8.96575	0.78440	9	
	1	0.99839	0.05669	0.99725	0.07410	0.99580	0.09150	1	
	2	1.99678	0.11338	1.99450	0.14821	1.99160	0.18300	2	
	3	2.99517	0.17007	2.99175	0.22232	2.98741	0.27450	3	
	4	3.99356	0.22677	3.98900	0.29643	3.98321	0.36600	4	
15'	5	4.99195	0.28346	4.98625	0.37054	4.97902	0.45750	5	45'
	6	5.99035	0.34015	5.98350	0.44465	5.97482	0.54900	6	
	7	6.98874	0.39684	6.98075	0.51875	6.97063	0.64051	7	
	8	7.98713	0.45354	7.97800	0.59286	7.96643	0.73201	8	
	9	8.98552	0.51023	8.97525	0.66697	8.96224	0.82351	9	
	1	0.99813	0.06104	0.99691	0.07845	0.99539	0.09584	1	
	2	1.99626	0.12209	1.99383	0.15691	1.99079	0.19169	2	
	3	2.99440	0.18314	2.99075	0.23537	2.98618	0.28753	3	
	4	3.99253	0.24419	3.98766	0.31383	3.98158	0.38338	4	
30'	5	4.99067	0.30524	4.98458	0.39229	4.97698	0.47922	5	30'
	6	5.98880	0.36629	5.98150	0.47075	5.97237	0.57507	6	
	7	6.98694	0.42733	6.97842	0.54921	6.96777	0.67092	7	
	8	7.98507	0.48838	7.97533	0.62767	7.96316	0.76676	8	
	9	8.98321	0.54943	8.97225	0.70613	8.95856	0.86261	9	
	1	9.99785	0.06540	0.99656	0.08280	0.99496	0.10018	1	
	2	1.99571	0.13080	1.99313	0.16561	1.98993	0.20037	2	
	3	2.99357	0.19620	2.98969	0.24842	2.98490	0.30056	3	
	4	3.99143	0.26161	3.98626	0.33123	3.97987	0.40075	4	
45'	5	4.98929	0.32701	4.98282	0.41404	4.97484	0.50094	5	15'
	6	5.98715	0.39241	5.97939	0.49684	5.96981	0.60112	6	
	7	6.98501	0.45782	6.97595	0.57965	6.96477	0.70131	7	
	8	7.98287	0.52322	7.97252	0.66246	7.95974	0.80150	8	
	9	8.98073	0.58862	8.96908	0.74527	8.95471	0.90169	9	
Minutes	Distance	Dep.	Lat.	Dep.	Lat.	Dep.	Lat.	Distance	Minutes
		86°		85°		84°			

Differences of Latitude and Departures—Continued.

Minutes.	Distance.	6°		7°		8°		Distance.	Minutes.
		Lat.	Dep.	Lat.	Dep.	Lat.	Dep.		
	1	0.99452	0.10452	0.99254	0.12186	0.99026	0.13917	1	
	2	1.98904	0.20905	1.98509	0.24373	1.98053	0.27834	2	
	3	2.98356	0.31358	2.97763	0.36560	2.97080	0.41751	3	
	4	3.97808	0.41811	3.97018	0.48747	3.96107	0.55669	4	
0′	5	4.97261	0.52264	4.96273	0.60934	4.95134	0.69586	5	60′
	6	5.96713	0.62717	5.95519	0.73121	5.94160	0.83503	6	
	7	6.96165	0.73169	6.94782	0.85308	6.93187	0.97421	7	
	8	7.95617	0.83622	7.94038	0.97495	7.92214	0.11338	8	
	9	8.95069	0.94075	8.93291	0.09682	8.91241	0.25255	9	
	1	0.99405	0.10886	0.99200	0.12619	0.98965	0.14349	1	
	2	1.98811	0.21773	1.98400	0.25239	1.97930	0.28698	2	
	3	2.98216	0.32660	2.97601	0.37859	2.96895	0.43047	3	
	4	3.97622	0.43546	3.96801	0.50479	3.95860	0.57397	4	
15′	5	4.97028	0.54433	4.96002	0.63099	4.94825	0.71746	5	45′
	6	5.96433	0.65320	5.95202	0.75719	5.93790	0.86095	6	
	7	6.95839	0.76206	6.94403	0.88339	6.92755	1.00444	7	
	8	7.95245	0.87093	7.93603	1.00959	7.91721	1.14794	8	
	9	8.94650	0.97980	8.92804	1.13579	8.90686	1.29143	9	
	1	0.99357	0.11320	0.99144	0.13052	0.98901	0.14780	1	
	2	1.98714	0.22640	1.98288	0.26105	1.97803	0.29561	2	
	3	2.98071	0.33960	2.97433	0.39157	2.96704	0.44342	3	
	4	3.97428	0.45281	3.96577	0.52210	3.95606	0.59123	4	
30′	5	4.96786	0.56601	4.95722	0.65263	4.94508	0.73904	5	30′
	6	5.96143	0.67921	5.94866	0.78315	5.93409	0.88685	6	
	7	6.95500	0.79242	6.94011	0.91368	6.92311	1.03466	7	
	8	7.94857	0.90562	7.93155	1.04420	7.91212	1.18247	8	
	9	8.94214	1.01882	8.92300	1.17473	8.90114	1.33028	9	
	1	0.99306	0.11753	0.99086	0.13485	0.98836	0.15212	1	
	2	1.98613	0.23507	1.98173	0.26970	1.97672	0.30424	2	
	3	2.97920	0.35261	2.97259	0.40455	2.96508	0.45637	3	
	4	3.97227	0.47014	3.96346	0.53940	3.95344	0.60849	4	
45′	5	4.96534	0.58768	4.95432	0.67425	4.94180	0.76061	5	15′
	6	5.95841	0.70522	5.94519	0.80910	5.93016	0.91274	6	
	7	6.95147	0.82276	6.93606	0.94395	6.91853	1.06486	7	
	8	7.94454	0.94029	7.92692	1.07880	7.90689	1.21698	8	
	9	8.93761	1.05783	8.91779	1.21365	8.89525	1.36911	9	
Minutes.	Distance.	Dep.	Lat.	Dep.	Lat.	Dep.	Lat.	Distance.	Minutes.
		83°		82°		81°			

Differences of Latitude and Departures—Continued.

Minutes.	Distance.	9° Lat.	9° Dep.	10° Lat.	10° Dep.	11° Lat.	11° Dep.	Distance.	Minutes.
	1	0.98768	0.15643	0.98480	0.17364	0.98162	0.19081	1	
	2	1.97537	0.31286	1.96961	0.34729	1.96325	0.38162	2	
	3	2.96306	0.46930	2.95442	0.52094	2.94488	0.57243	3	
	4	3.95075	0.62573	3.93923	0.69459	3.92650	0.76324	4	
0′	5	4.93844	0.78217	4.92403	0.86824	4.90813	0.95405	5	60′
	6	5.92612	0.93860	5.90884	1.04188	5.88976	1.14486	6	
	7	6.91381	1.09504	6.89365	1.21553	6.87139	1.33566	7	
	8	7.90150	1.25147	7.87846	1.38918	7.85301	1.52648	8	
	9	8.88919	1.40791	8.86327	1.56283	8.83464	1.71729	9	
	1	0.98699	0.16074	0.98404	0.17794	0.98078	0.19509	1	
	2	1.97399	0.32148	1.96808	0.35588	1.96157	0.39018	2	
	3	2.96098	0.48222	2.95212	0.53383	2.94235	0.58527	3	
	4	3.94798	0.64297	3.93616	0.71177	3.92314	0.78036	4	
15′	5	4.93498	0.80371	4.92020	0.88971	4.90392	0.97545	5	45′
	6	5.92197	0.96445	5.90424	1.06766	5.88471	1.17054	6	
	7	6.90897	1.12519	6.88828	1.24560	6.86549	1.36563	7	
	8	7.89597	1.28594	7.87232	1.42354	7.84628	1.56072	8	
	9	8.88296	1.44668	8.85636	1.60149	8.82706	1.75581	9	
	1	0.98628	0.16504	0.98325	0.18223	0.97992	0.19936	1	
	2	1.97257	0.33009	1.96650	0.36447	1.95984	0.39873	2	
	3	2.95885	0.49514	2.94976	0.54670	2.93977	0.59810	3	
	4	3.94514	0.66019	3.93301	0.72894	3.91969	0.79747	4	
30′	5	4.93142	0.82523	4.91627	0.91117	4.89962	0.99683	5	30′
	6	5.91771	0.99028	5.89952	1.09341	5.87954	1.19620	6	
	7	6.90399	1.15533	6.88278	1.27564	6.85947	1.39557	7	
	8	7.89028	1.32038	7.86603	1.45788	7.83939	1.59494	8	
	9	8.87657	1.48542	8.84929	1.64011	8.81932	1.79431	9	
	1	0.98555	0.16935	0.98245	0.18652	0.97904	0.20364	1	
	2	1.97111	0.33870	1.96490	0.37304	1.95809	0.40728	2	
	3	2.95666	0.50805	2.94735	0.55957	2.93713	0.61092	3	
	4	3.94222	0.67740	3.92980	0.74609	3.91618	0.81456	4	
45′	5	4.92778	0.84675	4.91225	0.93262	4.89522	1.01820	5	15′
	6	5.91333	1.01610	5.89470	1.11914	5.87427	1.22185	6	
	7	6.89889	1.18545	6.87715	1.30566	6.85331	1.42549	7	
	8	7.88444	1.35480	7.85960	1.49219	7.83236	1.62913	8	
	9	8.87000	1.52415	8.84205	1.67871	8.81140	1.83277	9	
Minutes.	Distance.	Dep.	Lat.	Dep.	Lat.	Dep.	Lat.	Distance.	Minutes.
		80°		79°		78°			

Differences of Latitude and Departures—Continued.

Minutes.	Distance.	12° Lat.	12° Dep.	13° Lat.	13° Dep.	14° Lat.	14° Dep.	Distance.	Minutes.
	1	0.97814	0.20791	0.97437	0.22495	0.97029	0.24192	1	
	2	1.95629	0.41582	1.94874	0.44990	1.94059	0.48384	2	
	3	2.93444	0.62373	2.92311	0.67485	2.91088	0.72576	3	
	4	3.91259	0.83164	3.89748	0.89980	3.88118	0.96768	4	
0′	5	4.89073	1.03955	4.87185	1.12475	4.85147	1.20961	5	60′
	6	5.86888	1.24747	5.84622	1.34970	5.82177	1.45153	6	
	7	6.84703	1.45538	6.82059	1.57465	6.79206	1.69345	7	
	8	7.82518	1.66329	7.79496	1.79960	7.76236	1.93537	8	
	9	8.80332	1.87120	8.76933	2.02455	8.73266	2.17729	9	
	1	0.97723	0.21217	0.97337	0.22920	0.96923	0.24615	1	
	2	1.95446	0.42435	1.94675	0.45840	1.93846	0.49230	2	
	3	2.93169	0.63653	2.92013	0.68760	2.90769	0.73845	3	
	4	3.90892	0.84871	3.89351	0.91680	3.87692	0.98461	4	
15′	5	4.88615	1.06088	4.86689	1.14600	4.84615	1.23076	5	45′
	6	5.86338	1.27306	5.84027	1.37520	5.81538	1.47691	6	
	7	6.84061	1.48524	6.81365	1.60440	6.78461	1.72307	7	
	8	7.81784	1.69742	7.78703	1.83360	7.75384	1.96922	8	
	9	8.79507	1.90959	8.76041	2.06280	8.72307	2.21537	9	
	1	0.97629	0.21644	0.97237	0.23344	0.96814	0.25038	1	
	2	1.95259	0.43288	1.94474	0.46689	1.93629	0.50076	2	
	3	2.92888	0.64932	2.91711	0.70033	2.90444	0.75114	3	
	4	3.90518	0.86576	3.88948	0.93378	3.87259	1.00152	4	
30′	5	4.88148	1.08220	4.86185	1.16722	4.84073	1.25190	5	30′
	6	5.85777	1.29864	5.83422	1.40067	5.80888	1.50228	6	
	7	6.83407	1.51508	6.80659	1.63411	6.77703	1.75266	7	
	8	7.81036	1.73152	7.77896	1.86756	7.74518	2.00304	8	
	9	8.78666	1.94796	8.75133	2.10100	8.71332	2.25342	9	
	1	0.97534	0.22069	0.97134	0.23768	0.96704	0.25460	1	
	2	1.95068	0.44139	1.94268	0.47537	1.93409	0.50920	2	
	3	2.92602	0.66209	2.91402	0.71305	2.90113	0.76380	3	
	4	3.90136	0.88278	3.88536	0.95074	3.86818	1.01840	4	
45′	5	4.87671	1.10348	4.85671	1.18843	4.83523	1.27301	5	15′
	6	5.85205	1.32418	5.82805	1.42611	5.80227	1.52761	6	
	7	6.82739	1.54488	6.79939	1.66380	6.76932	1.78221	7	
	8	7.80273	1.76557	7.77073	1.90148	7.73636	2.03681	8	
	9	8.77808	1.98627	8.74207	2.13917	8.70341	2.29141	9	
Minutes.	Distance.	77° Dep.	77° Lat.	76° Dep.	76° Lat.	75° Dep.	75° Lat.	Distance.	Minutes.

Differences of Latitude and Departures—Continued.

Minutes.	Dist.	15° Lat.	15° Dep.	16° Lat.	16° Dep.	17° Lat.	17° Dep.	Distance.	Minutes.
0′	1	0.96592	0.25881	0.96126	0.27563	0.95630	0.29237	1	60′
	2	1.93185	0.51763	1.92252	0.55127	1.91260	0.58474	2	
	3	2.89777	0.77645	2.88378	0.82691	2.86891	0.87711	3	
	4	3.86370	1.03527	3.84504	1.10254	3.82521	1.16948	4	
	5	4.82962	1.29409	4.80630	1.37818	4.78152	1.46185	5	
	6	5.79555	1.55291	5.76757	1.65382	5.73782	1.75423	6	
	7	6.76148	1.81173	6.72883	1.92946	6.69413	2.04660	7	
	8	7.72740	2.07055	7.69009	2.20509	7.65043	2.33897	8	
	9	8.69333	2.32937	8.65135	2.48073	8.60674	2.63134	9	
15′	1	0.96478	0.26303	0.96005	0.27982	0.95502	0.29654	1	45′
	2	1.92957	0.52606	1.92010	0.55965	1.91004	0.59308	2	
	3	2.89436	0.78909	2.88015	0.83948	2.86506	0.88962	3	
	4	3.85914	1.05212	3.84020	1.11931	3.82008	1.18616	4	
	5	4.82393	1.31515	4.80025	1.39914	4.77510	1.48270	5	
	6	5.78872	1.57818	5.76030	1.67897	5.73012	1.77924	6	
	7	6.75351	1.84121	6.72035	1.95880	6.68514	2.07579	7	
	8	7.71829	2.10424	7.68040	2.23863	7.64016	2.37233	8	
	9	8.68308	2.36728	8.64045	2.51846	8.59518	2.66887	9	
30′	1	0.96363	0.26723	0.95882	0.28401	0.95371	0.30070	1	30′
	2	1.92726	0.53447	1.91764	0.56803	1.90743	0.60141	2	
	3	2.89089	0.80171	2.87646	0.85204	2.86115	0.90211	3	
	4	3.85452	1.06895	3.83528	1.13606	3.81486	1.20282	4	
	5	4.81815	1.33619	4.79410	1.42007	4.76858	1.50352	5	
	6	5.78178	1.60343	5.75292	1.70409	5.72230	1.80423	6	
	7	6.74541	1.87066	6.71174	1.98810	6.67601	2.10494	7	
	8	7.70904	2.13790	7.67056	2.27212	7.62973	2.40564	8	
	9	8.67267	2.40514	8.62938	2.55613	8.58345	2.70635	9	
45′	1	0.96245	0.27144	0.95757	0.28819	0.95239	0.30486	1	15′
	2	1.92491	0.54288	1.91514	0.57639	1.90479	0.60972	2	
	3	2.88736	0.81432	2.87271	0.86458	2.85718	0.91459	3	
	4	3.84982	1.08576	3.83028	1.15278	3.80958	1.21945	4	
	5	4.81227	1.35720	4.78785	1.44098	4.76197	1.52432	5	
	6	5.77473	1.62864	5.74542	1.72917	5.71437	1.82918	6	
	7	6.73718	1.90008	6.70299	2.01737	6.66677	2.13405	7	
	8	7.69964	2.17152	7.66057	2.30557	7.61916	2.43891	8	
	9	8.66209	2.44296	8.61814	2.59376	8.57156	2.74377	9	
Minutes.	Dist.	Dep.	Lat.	Dep.	Lat.	Dep.	Lat.	Distance.	Minutes.
		74°		73°		72°			

Differences of Latitude and Departures—Continued.

Minutes	Distance	18° Lat.	18° Dep.	19° Lat.	19° Dep.	20° Lat.	20° Dep.	Distance	Minutes
	1	0.95105	0.30901	0.94551	0.32556	0.93969	0.34202	1	
	2	1.90211	0.61803	1.89103	0.65113	1.87938	0.68404	2	
	3	2.85316	0.92705	2.83655	0.97670	2.81907	1.02606	3	
	4	3.80422	1.23606	3.78207	1.30227	3.75877	1.36808	4	
0′	5	4.75528	1.54508	4.72759	1.62784	4.69846	1.71010	5	60′
	6	5.70633	1.85410	5.67311	1.95340	5.63815	2.05212	6	
	7	6.65739	2.16311	6.61863	2.27897	6.57784	2.39414	7	
	8	7.60845	2.47213	7.56414	2.60454	7.51754	2.73616	8	
	9	8.55950	2.78115	8.50966	2.93011	8.45723	3.07818	9	
	1	0.94969	0.31316	0.94408	0.32969	0.93819	0.34611	1	
	2	1.89939	0.62632	1.88817	0.65938	1.87638	0.69223	2	
	3	2.84909	0.93949	2.83226	0.98907	2.81457	1.03835	3	
	4	3.79879	1.25265	3.77635	1.31876	3.75276	1.38446	4	
15′	5	4.74849	1.56581	4.72044	1.64845	4.69095	1.73058	5	45′
	6	5.69819	1.87898	5.66453	1.97814	5.62914	2.07670	6	
	7	6.64789	2.19214	6.60862	2.30783	6.56733	2.44281	7	
	8	7.59759	2.50531	7.55271	2.63752	7.50553	2.76893	8	
	9	8.54729	2.81847	8.49680	2.96721	8.44372	3.11505	9	
	1	0.94832	0.31730	0.94264	0.33380	0.93667	0.35020	1	
	2	1.89664	0.63460	1.88528	0.66761	1.87334	0.70041	2	
	3	2.84497	0.95191	2.82792	1.00142	2.81001	1.05062	3	
	4	3.79329	1.26921	3.77056	1.33522	3.74668	1.40082	4	
30′	5	4.74161	1.58652	4.71320	1.66903	4.68336	1.75103	5	30′
	6	5.68994	1.90382	5.65584	2.00284	5.62003	2.10124	6	
	7	6.63826	2.22113	6.59849	2.33664	6.55670	2.45145	7	
	8	7.58658	2.53843	7.54113	2.67045	7.49337	2.80165	8	
	9	8.53491	2.85574	8.48377	3.00426	8.43004	3.15186	9	
	1	0.94693	0.32143	0.94117	0.33791	0.93513	0.35429	1	
	2	1.89386	0.64287	1.88235	0.67583	1.87027	0.70858	2	
	3	2.84079	0.96431	2.82352	1.01375	2.80540	1.06287	3	
	4	3.78772	1.28575	3.76470	1.35166	3.74054	1.41716	4	
45′	5	4.73465	1.60719	4.70588	1.68958	4.67567	1.77145	5	15′
	6	5.68158	1.92863	5.64705	2.02750	5.61081	2.12574	6	
	7	6.62851	2.25007	6.58823	2.36541	6.54594	2.48003	7	
	8	7.57544	2.57151	7.52940	2.70333	7.48108	2.83432	8	
	9	8.52237	2.89295	8.47058	3.04125	8.41621	3.18861	9	
Minutes / Distance		Dep.	Lat.	Dep.	Lat.	Dep.	Dat.	Distance	Minutes
		71°		70°		69°			

Differences of Latitude and Departures—Continued.

Minutes.	Distance.	21° Lat.	21° Dep.	22° Lat.	22° Dep.	23° Lat.	23° Dep.	Distance.	Minutes.
0′	1	0.93358	0.35836	0.92718	0.37460	0.92050	0.39073	1	60′
	2	1.86716	0.71673	1.85436	0.74921	1.84100	0.78146	1	
	3	2.80074	1.07510	2.78155	1.12381	2.76151	1.17219	3	
	4	3.73432	1.43347	3.70873	1.49842	3.68201	1.56292	4	
	5	4.66790	1.79183	4.63591	1.87303	4.60252	1.95365	5	
	6	5.60148	2.15020	5.56310	2.24763	5.52302	2.34438	6	
	7	6.53506	2.50857	6.49028	2.62224	6.44353	2.73511	7	
	8	7.46864	2.86694	7.41747	2.99685	7.36403	3.12584	8	
	9	8.40222	3.22531	8.34465	3.37145	8.28454	3.51657	9	
15′	1	0.93200	0.36243	0.92554	0.37864	0.91879	0.39474	1	45′
	2	1.86401	0.72487	1.85108	0.75729	1.83758	0.78948	2	
	3	2.79602	1.08731	2.77662	1.13594	2.75637	1.18423	3	
	4	3.72803	1.44975	3.70216	1.51459	3.67516	1.57897	4	
	5	4.66004	1.81219	4.62770	1.89324	4.59395	1.97372	5	
	6	5.59204	2.17462	5.55324	2.27189	5.51274	2.36846	6	
	7	6.52405	2.53706	6.47878	2.65054	6.43153	2.76320	7	
	8	7.45606	2.89950	7.40432	3.02918	7.35032	3.15795	8	
	9	8.38807	3.26194	8.32986	3.40783	8.26912	3.55269	9	
30′	1	0.93041	0.36650	0.92388	0.38268	0.91706	0.39874	1	30′
	2	1.86083	0.73300	1.84776	0.76536	1.83412	0.79749	2	
	3	2.79125	1.09950	2.77164	1.14805	2.75118	1.19624	3	
	4	3.72167	1.46600	3.69552	1.53073	3.66824	1.59499	4	
	5	4.65208	1.83250	4.61940	1.91341	4.58530	1.99374	5	
	6	5.58250	2.19900	5.54328	2.29610	5.50236	2.39249	6	
	7	6.51292	2.56550	6.46716	2.67878	6.41942	2.79124	7	
	8	7.44334	2.93200	7.39104	3.06146	7.33648	3.18999	8	
	9	8.37375	3.29851	8.31492	3.44415	8.25354	3.58874	9	
45′	1	0.92881	0.37055	0.92220	0.38671	0.91531	0.40274	1	15′
	2	1.85762	0.74111	1.84440	0.77342	1.83062	0.80549	2	
	3	2.78643	1.11167	2.76660	1.16013	2.74593	1.20824	3	
	4	3.71524	1.48222	3.68880	1.54684	3.66124	1.61098	4	
	5	4.64405	1.85278	4.61100	1.93355	4.57655	2.01373	5	
	6	5.57286	2.22334	5.53320	2.32026	5.49186	2.41648	6	
	7	6.50167	2.59390	6.45540	2.70697	6.40718	2.81922	7	
	8	7.43048	2.96445	7.37760	3.09368	7.32249	3.22197	8	
	9	8.35929	3.33501	8.29980	3.48039	8.23780	3.62472	9	
Minutes.	Distance.	68° Dep.	Lat.	67° Dep.	Lat.	66° Dep.	Lat.	Distance.	Minutes.

Differences of *Latitude* and *Departures*—Continued.

Minutes.	Distance.	24° Lat.	24° Dep.	25° Lat.	25° Dep.	26° Lat.	26° Dep.	Distance.	Minutes.
	1	0.91354	0.40673	0.90630	0.42261	0.89879	0.43837	1	
	2	1.82709	0.81347	1.81261	0.84523	1.79758	0.87674	2	
	3	2.74063	1.22020	2.71892	1.26785	2.69638	1.31511	3	
	4	3.65418	1.62694	3.62523	1.69047	3.59517	1.75348	4	
0′	5	4.56772	2.03368	4.53153	2.11309	4.49397	2.19185	5	60′
	6	5.48127	2.44041	5.43784	2.53570	5.39276	2.63022	6	
	7	6.39481	2.84715	6.34415	2.95832	6.29155	3.06859	7	
	8	7.30836	3.25389	7.25046	3.38094	7.19035	3.50696	8	
	9	8.22190	3.66062	8.15677	3.80356	8.08914	3.94533	9	
	1	0.91176	0.41071	0.90445	0.42656	0.89687	0.44228	1	
	2	1.82352	0.82143	1.80891	0.85313	1.79374	0.88457	2	
	3	2.73528	1.23215	2.71336	1.27970	2.69061	1.32686	3	
	4	3.64704	1.64287	3.61782	1.70627	3.58749	1.76915	4	
15′	5	4.55881	2.05359	4.52227	2.13284	4.48436	2.21144	5	45′
	6	5.47057	2.46431	5.42673	2.55941	5.38123	2.65373	6	
	7	6.38233	2.87503	6.33118	2.98598	6.27810	3.09602	7	
	8	7.29409	3.28575	7.23564	3.41254	7.17498	3.53830	8	
	9	8.20585	3.69647	8.14009	3.83911	8.07185	3.98059	9	
	1	0.90996	0.41469	0.90258	0.43051	0.89493	0.44619	1	
	2	1.81992	0.82938	1.80517	0.86102	1.78986	0.89239	2	
	3	2.72988	1.24407	2.70775	1.29153	2.68480	1.33859	3	
	4	3.63984	1.65877	3.61034	1.72204	3.57973	1.78479	4	
30′	5	4.54980	2.07346	4.51292	2.15255	4.47467	2.23098	5	30′
	6	5.45976	2.48815	5.41551	2.58306	5.36960	2.67718	6	
	7	6.36972	2.90285	6.31809	3.01357	6.26454	3.12338	7	
	8	7.27969	3.31754	7.22068	3.44408	7.15947	3.56958	8	
	9	8.18965	3.73223	8.12326	3.87459	8.05440	4.01578	9	
	1	0.90814	0.41866	0.90069	0.43444	0.89297	0.45009	1	
	2	1.81628	0.83732	1.80139	0.86889	1.78595	0.90019	2	
	3	2.72442	1.25598	2.70209	1.30333	2.67893	1.35029	3	
	4	3.63257	1.67464	3.60279	1.73778	3.57191	1.80039	4	
45′	5	4.54071	2.09330	4.50349	2.17222	4.46489	2.25049	5	15′
	6	5.44885	2.51196	5.40418	2.60667	5.35787	2.70059	6	
	7	6.35700	2.93062	6.30488	3.04111	6.25085	3.15068	7	
	8	7.26514	3.34928	7.20558	3.47556	7.14383	3.60078	8	
	9	8.17328	3.76794	8.10628	3.91000	8.03681	4.05088	9	
Minutes.	Distance.	65° Dep.	65° Lat.	64° Dep.	64° Lat.	63° Dep.	63° Lat.	Distance.	Minutes.

Differences of Latitude and Departures—Continued.

Minutes.	Distance.	27° Lat.	27° Dep.	28° Lat.	28° Dep.	29° Lat.	29° Dep.	Distance.	Minutes.
0′	1	0.89100	0.45399	0.88294	0.46947	0.87462	0.48481	1	60′
	2	1.78201	0.90798	1.76589	0.93894	1.74924	0.96962	2	
	3	2.67301	1.36197	2.64884	1.40841	2.62386	1.45443	3	
	4	3.56402	1.81596	3.53179	1.87788	3.49848	1.93924	4	
	5	4.45503	2.26995	4.41473	2.34735	4.37310	2.42405	5	
	6	5.34603	2.72394	5.29768	2.81682	5.24772	2.90886	6	
	7	6.23704	3.17793	6.18063	3.28630	6.12234	3.39367	7	
	8	7.12805	3.63193	7.06358	3.75577	6.99696	3.87848	8	
	9	8.01905	4.08591	7.94652	4.22524	7.87156	4.36329	9	
15′	1	0.88901	0.45787	0.88089	0.47332	0.87249	0.48862	1	45′
	2	1.77803	0.91574	1.76178	0.94664	1.74499	0.97724	2	
	3	2.66705	1.37362	2.64267	1.41996	2.61748	1.46566	3	
	4	3.55606	1.83149	3.52356	1.89328	3.48998	1.95448	4	
	5	4.44508	2.28937	4.40445	2.36660	4.36248	2.44310	5	
	6	5.33410	2.74724	5.28534	2.83992	5.23497	2.93172	6	
	7	6.22311	3.20511	6.16623	3.31324	6.10747	3.42034	7	
	8	7.11213	3.66299	7.04712	3.78656	6.97996	3.90896	8	
	9	8.00115	4.12086	7.92801	4.25988	7.85246	4.39759	9	
30′	1	0.88701	0.46174	0.87881	0.47715	0.87035	0.49242	1	30′
	2	1.77402	0.92349	1.75763	0.95431	1.74071	0.98484	2	
	3	2.66103	1.38524	2.63645	1.43147	2.61106	1.47727	3	
	4	3.54804	1.84699	3.51526	1.90863	3.48142	1.96969	4	
	5	4.43505	2.30874	4.39408	2.38579	4.35177	2.46211	5	
	6	5.32206	2.77049	5.27290	2.86295	5.22213	2.95454	6	
	7	6.20907	3.23224	6.15171	3.34011	6.09248	3.44696	7	
	8	7.09608	3.69398	7.03053	3.81727	6.96284	3.93938	8	
	9	7.98309	4.15573	7.90935	4.29442	7.83320	4.43181	9	
45′	1	0.88498	0.46561	0.87672	0.48098	0.86819	0.49621	1	15′
	2	1.76997	0.93122	1.75345	0.96197	1.73639	0.99243	2	
	3	2.65496	1.39684	2.63018	1.44296	2.60459	1.48864	3	
	4	3.53995	1.86245	3.50690	1.92395	3.47279	1.98486	4	
	5	4.42493	2.32807	4.38363	2.40494	4.34099	2.48108	5	
	6	5.30992	2.79368	5.26036	2.88593	5.20919	2.97729	6	
	7	6.19491	3.25930	6.13708	3.36692	6.07739	3.47351	7	
	8	7.07990	3.72491	7.01381	3.84791	6.94559	3.96973	8	
	9	7.96488	4.19053	7.89054	4.32889	7.81378	4.46594	9	
Minutes.	Distance.	Dep.	Lat.	Dep.	Lat.	Dep.	Lat.	Distance.	Minutes.
		62°		61°		60°			

Differences of Latitude and Departures—Continued.

Minutes.	Distance.	30° Lat.	30° Dep.	31° Lat.	31° Dep.	32° Lat.	32° Dep.	Distance.	Minutes.
0′	1	0.86602	0.50000	0.85716	0.51503	0.84804	0.52991	1	60′
	2	1.73205	1.00000	1.71433	1.03007	1.69609	1.05983	2	
	3	2.59807	1.50000	2.57150	1.54511	2.54414	1.58975	3	
	4	3.46410	2.00000	3.42866	2.06015	3.39219	2.11967	4	
	5	4.33012	2.50000	4.28583	2.57519	4.24024	2.64959	5	
	6	5.19615	3.00000	5.14300	3.09022	5.08828	3.17951	6	
	7	6.06217	3.50000	6.00017	3.60526	5.93633	3.70943	7	
	8	6.92820	4.00000	6.85733	4.12030	6.78438	4.23935	8	
	9	7.79422	4.50000	7.71450	4.63534	7.63243	4.76927	9	
15′	1	0.86383	0.50377	0.85491	0.51877	0.84572	0.53361	1	45′
	2	1.72767	1.00754	1.70982	1.03754	1.69145	1.06722	2	
	3	2.59150	1.51132	2.56473	1.55631	2.53718	1.60084	3	
	4	3.45534	2.01509	3.41964	2.07509	3.38291	2.13445	4	
	5	4.31917	2.51887	4.27456	2.59386	4.22863	2.66807	5	
	6	5.18301	3.02264	5.12947	3.11263	5.07436	3.20168	6	
	7	6.04684	3.52641	5.98438	3.63141	5.92009	3.73530	7	
	8	6.91068	4.03019	6.83929	4.15018	6.76582	4.26891	8	
	9	7.77451	4.53396	7.69420	4.66895	7.61155	4.80253	9	
30′	1	0.86162	0.50753	0.85264	0.52249	0.84339	0.53730	1	30′
	2	1.72325	1.01507	1.70528	1.04499	1.68678	1.07460	2	
	3	2.58488	1.52261	2.55792	1.56749	2.53017	1.61190	3	
	4	3.44651	2.03015	3.41056	2.08999	3.37356	2.14920	4	
	5	4.30814	2.53769	4.26320	2.61249	4.21695	2.68650	5	
	6	5.16977	3.04523	5.11584	3.13499	5.06034	3.22380	6	
	7	6.03140	3.55276	5.96948	3.65749	5.90373	3.76110	7	
	8	6.89303	4.06030	6.82112	4.17998	6.74713	4.29840	8	
	9	7.75466	4.56784	7.67376	4.70248	7.59052	4.83570	9	
45′	1	0.85940	0.51129	0.85035	0.52621	0.84103	0.54097	1	15′
	2	1.71881	1.02258	1.70070	1.05242	1.68207	1.08194	2	
	3	2.57821	1.53387	2.55105	1.57864	2.52311	1.62292	3	
	4	3.43762	2.04517	3.40140	2.10485	3.36415	2.16389	4	
	5	4.29703	2.55646	4.25176	2.63107	4.20519	2.70487	5	
	6	5.15643	3.06775	5.10211	3.15728	5.04623	3.24584	6	
	7	6.01584	3.57905	5.95246	3.68349	5.88827	3.78682	7	
	8	6.87525	4.09034	6.80281	4.20971	6.72831	4.32779	8	
	9	7.73465	4.60163	7.65316	4.73592	7.56935	4.86877	9	
Minutes.	Distance.	Dep.	Lat.	Dep.	Lat.	Dep.	Lat.	Distance.	Minutes.
		59°		58°		57°			

Differences of Latitude and Departures—Continued.

Minutes.	Distance.	33°		34°		35°		Distance.	Minutes.
		Lat.	Dep.	Lat.	Dep.	Lat.	Dep.		
0′	1	0.83867	0.54463	0.82903	0.55919	0.81915	0.57357	1	60′
	2	1.67734	1.08927	1.65807	1.11838	1.63830	1.14715	2	
	3	2.51601	1.63391	2.48711	1.67757	2.45745	1.72072	3	
	4	3.35468	2.17855	3.31615	2.23677	3.27660	2.29430	4	
	5	4.19335	2.72319	4.14518	2.79596	4.09576	2.86788	5	
	6	5.03202	3.26783	4.97422	3.35515	4.91491	3.44145	6	
	7	5.87069	3.81247	5.80326	3.91435	5.73406	4.01503	7	
	8	6.70936	4.35711	6.63230	4.47354	6.55321	4.58861	8	
	9	7.54803	4.90175	7.46133	5.03273	7.37236	5.16218	9	
15′	1	0.83628	0.54829	0.82659	0.56280	0.81664	0.57714	1	45′
	2	1.67257	1.09658	1.65318	1.12560	1.63328	1.15429	2	
	3	2.50885	1.64487	2.47977	1.68841	2.44992	1.73143	3	
	4	3.34514	2.19317	3.30636	2.25121	3.26656	2.30858	4	
	5	4.18143	2.74146	4.13295	2.81402	4.08320	2.88572	5	
	6	5.01771	3.28975	4.95954	3.37682	4.89984	3.46287	6	
	7	5.85400	3.83805	5.78613	3.93963	5.71649	4.04001	7	
	8	6.69028	4.38634	6.61272	4.50243	6.53313	4.61716	8	
	9	7.52657	4.93463	7.43931	5.06524	7.34977	5.19430	9	
30′	1	0.83388	0.55193	0.82412	0.56640	0.81411	0.58070	1	30′
	2	1.66777	1.10387	1.64825	1.13281	1.62823	1.16140	2	
	3	2.50165	1.65581	2.47237	1.69921	2.44234	1.74210	3	
	4	3.33554	2.20774	3.29650	2.26562	3.25646	2.32281	4	
	5	4.16942	2.75968	4.12063	2.83203	4.07057	2.90351	5	
	6	5.00331	3.31162	4.94475	3.39843	4.88469	3.48421	6	
	7	5.83720	3.86355	5.76888	3.96484	5.69880	4.06492	7	
	8	6.67108	4.41549	6.59300	4.53124	6.51292	4.64562	8	
	9	7.50497	4.96743	7.41713	5.09765	7.32703	5.22632	9	
45′	1	0.83147	0.55557	0.82164	0.56999	0.81157	0.58425	1	15′
	2	1.66294	1.11114	1.64329	1.13999	1.62314	1.16850	2	
	3	2.49441	1.66671	2.46494	1.70999	2.43472	1.75275	3	
	4	3.32588	2.22228	3.28658	2.27998	3.24629	2.33700	4	
	5	4.15735	2.77785	4.10823	2.84998	4.05787	2.92125	5	
	6	4.98882	3.33342	4.92988	3.41998	4.86944	3.50550	6	
	7	5.82029	3.88899	5.75152	3.98997	5.68101	4.08975	7	
	8	6.65176	4.44456	6.57317	4.55997	6.49260	4.67400	8	
	9	7.48323	5.00013	7.39482	5.12997	7.30416	5.25825	9	
Minutes.	Distance.	Dep.	Lat.	Dep.	Lat.	Dep.	Lat.	Distance.	Minutes.
		56°		55°		54°			

Differences of Latitude and Departures—Continued.

Minutes.	Distance.	36° Lat.	36° Dep.	37° Lat.	37° Dep.	38° Lat.	38° Dep.	Distance.	Minutes.
	1	0.80901	0.58778	0.79863	0.60181	0.78801	0.61566	1	
	2	1.61803	1.17557	1.59727	1.20363	1.57602	1.23132	2	
	3	2.42705	1.76335	2.39590	1.80544	2.36403	1.84698	3	
	4	3.23606	2.35114	3.19454	2.40726	3.15204	2.46264	4	
0′	5	4.04508	2.93892	3.99317	3.00907	3.94005	3.07830	5	60′
	6	4.85410	3.52671	4.79181	3.61089	4.72806	3.69396	6	
	7	5.66311	4.11449	5.59044	4.21270	5.51607	4.30963	7	
	8	6.47213	4.70228	6.38908	4.81452	6.30408	4.92529	8	
	9	7.28115	5.29006	7.18771	5.41633	7.09209	5.54095	9	
	1	0.80644	0.59130	0.79600	0.60529	1.78531	0.61909	1	
	2	1.61288	1.18261	1.59200	1.21058	1.57063	1.23818	2	
	3	2.41933	1.77392	2.38800	1.81588	2.35595	1.85728	3	
	4	3.22577	2.36523	3.18400	2.42117	3.14126	2.47637	4	
15′	5	4.03222	2.95654	3.98001	3.02647	3.92658	3.09547	5	45′
	6	4.83866	3.54785	4.77601	3.63176	4.71190	3.71456	6	
	7	5.64511	4.13916	5.57201	4.23705	5.49721	4.33365	7	
	8	6.45155	4.73047	6.36801	4.84235	6.28253	4.95275	8	
	9	7.25800	5.32178	7.16401	5.44764	7.06785	5.57184	9	
	1	0.80385	0.59482	0.79335	0.60876	0.78260	0.62251	1	
	2	1.60771	1.18964	1.58670	1.21752	1.56521	1.24502	2	
	3	2.41157	1.78446	2.38005	1.82628	2.34782	1.86754	3	
	4	3.21542	2.37929	3.17341	2.43504	3.13043	2.49005	4	
30′	5	4.01928	2.97411	3.96676	3.04380	3.91304	3.11257	5	30′
	6	4.82314	3.56893	4.76011	3.65256	4.69564	3.73508	6	
	7	5.62699	4.16375	5.55347	4.26132	5.47825	4.35760	7	
	8	6.43085	4.75858	6.34682	4.87009	6.26086	4.98011	8	
	9	7.23471	5.35340	7.14017	5.47885	7.04347	5.60263	9	
	1	0.80125	0.59832	0.79068	0.61221	0.77988	0.62592	1	
	2	1.60250	1.19664	1.58137	1.22443	1.55946	1.25184	2	
	3	2.40376	1.79497	2.37206	1.83665	2.33965	1.87777	3	
	4	3.20501	2.39329	3.16275	2.44886	3.11953	2.50369	4	
45′	5	4.00626	2.99162	3.95344	3.06108	3.89942	3.12961	5	15′
	6	4.80752	3.58994	4.74413	3.67330	4.67930	3.75554	6	
	7	5.60877	4.18827	5.53482	4.28552	5.45919	4.38146	7	
	8	6.41003	4.78659	6.32551	4.89773	6.23907	5.00738	8	
	9	7.21128	5.38492	7.11620	5.50995	7.01896	5.63331	9	
Minutes.	Distance.	Dep.	Lat.	Dep.	Lat.	Dep.	Lat.	Distance.	Minutes.
		53°		52°		51°			

Differences of Latitude and Departures—Continued.

Minutes.	Distance.	39°		40°		41°		Distance.	Minutes.
		Lat.	Dep.	Lat.	Dep.	Lat.	Dep.		
0′	1	0.77714	0.62932	0.76604	0.64278	0.75470	0.65605	1	60′
	2	1.55429	1.25864	1.53208	1.28557	1.50941	1.31211	2	
	3	2.33143	1.88796	2.29813	1.92836	2.26412	1.96817	3	
	4	3.10858	2.51728	3.06417	2.57115	3.01883	2.62423	4	
	5	3.88573	3.14660	3.83022	3.21393	3.77354	3.28029	5	
	6	4.66287	3.77592	4.59626	3.85672	4.52825	3.93635	6	
	7	5.44002	4.40524	5.36231	4.49951	5.28296	4.59241	7	
	8	6.21716	5.03456	6.12835	5.14230	6.03767	5.24847	8	
	9	6.99431	5.66388	6.89439	5.78508	6.79238	5.90453	9	
15′	1	0.77439	0.63270	0.76323	0.64612	0.75184	0.65934	1	45′
	2	1.54878	1.26541	1.52646	1.29224	1.50368	1.31869	2	
	3	2.32317	1.89811	2.28969	1.93837	2.25552	1.97803	3	
	4	3.09757	2.53082	3.05293	2.58449	3.00736	2.63738	4	
	5	3.87196	3.16352	3.81616	3.23062	3.75920	3.29672	5	
	6	4.64635	3.79623	4.57939	3.87674	4.51104	3.95607	6	
	7	5.42074	4.42893	5.34262	4.52286	5.26288	4.61542	7	
	8	6.19514	5.06164	6.10586	5.16899	6.01472	5.27476	8	
	9	6.96953	5.69434	6.86909	5.81511	6.76656	5.93411	9	
30′	1	0.77162	0.63607	0.76040	0.64944	0.74895	0.66262	1	30′
	2	1.54324	1.27215	1.52081	1.29889	1.49791	1.32524	2	
	3	2.31487	1.90823	2.28121	1.94834	2.24686	1.98786	3	
	4	3.08649	2.54431	3.04162	2.59779	2.99582	2.65048	4	
	5	3.85812	3.18039	3.80203	3.24724	3.74477	3.31310	5	
	6	4.62974	3.81646	4.56243	3.89668	4.49373	3.97572	6	
	7	5.40137	4.45254	5.32284	4.54613	5.24268	4.63834	7	
	8	6.17299	5.08862	6.08324	5.19558	5.99164	5.30096	8	
	9	6.94462	5.72470	6.84365	5.84503	6.74060	5.96358	9	
45′	1	0.76884	0.63943	0.75756	0.65276	0.74605	0.66588	1	15′
	2	1.53768	1.27887	1.51513	1.30552	1.49211	1.33176	2	
	3	2.30652	1.91831	2.27269	1.95828	2.23817	1.99764	3	
	4	3.07536	2.55775	3.03026	2.61104	2.98422	2.66352	4	
	5	3.84420	3.19719	3.78782	2.26380	3.73028	3.32940	5	
	6	4.61305	3.83663	4.54539	3.91656	4.47634	3.99529	6	
	7	5.38189	4.47607	5.30295	4.56932	5.22240	4.66117	7	
	8	6.15073	5.11551	6.06052	5.22208	5.96845	5.32705	8	
	9	6.91957	5.75495	6.81808	5.87484	6.71451	5.99293	9	
Minutes.	Distance.	Dep.	Lat.	Dep.	Lat.	Dep.	Lat.	Distance.	Minutes.
		50°		49°		48°			

Differences of Latitude and Departures—Continued.

Minutes	Distance	42° Lat.	42° Dep.	43° Lat.	43° Dep.	44° Lat.	44° Dep.	Distance	Minutes
0′	1	0.74314	0.66913	0.73135	0.68199	0.71933	0.69465	1	60′
	2	1.48628	1.33826	1.46270	1.36399	1.43867	1.38931	2	
	3	2.22943	2.00739	2.19406	2.04599	2.15801	2.08397	3	
	4	2.97257	2.67652	2.92541	2.72799	2.87735	2.77863	4	
	5	3.71572	3.34565	3.65676	3.40999	3.59669	3.47329	5	
	6	4.45886	4.01478	4.38812	4.09199	4.31603	4.16795	6	
	7	5.20201	4.68391	5.11947	4.77398	5.03537	4.86260	7	
	8	5.94515	5.35304	5.85082	5.45598	5.75471	5.55726	8	
	9	6.68830	6.02217	6.58218	6.13798	6.47405	6.25192	9	
15′	1	0.74021	0.67236	0.72837	1.68518	0.71630	0.69779	1	45′
	2	1.48043	1.34473	1.45674	0.37036	1.43260	1.39558	2	
	3	2.22065	2.01710	2.18511	2.05554	2.14890	2.09337	3	
	4	2.96087	2.68946	2.91348	2.74073	2.86520	2.79116	4	
	5	3.70109	3.36183	3.64185	3.42591	3.58151	3.48895	5	
	6	4.44130	4.03420	4.37022	4.11109	4.29781	4.18674	6	
	7	5.18152	4.70656	5.09859	4.79628	5.01411	4.88453	7	
	8	5.92174	5.37893	5.82696	5.48146	5.73041	5.58232	8	
	9	6.66196	6.05130	6.55533	6.16664	6.44671	6.28011	9	
30′	1	0.73727	0.67559	0.72537	0.68835	0.71325	0.70090	1	30′
	2	1.47455	1.35118	1.45074	1.37670	1.42650	1.40181	2	
	3	2.21183	2.02677	2.17612	2.06506	2.13975	2.10272	3	
	4	2.94910	2.70236	2.90149	2.75341	2.85300	2.80363	4	
	5	3.68638	3.37795	3.62687	3.44177	3.56625	3.50454	5	
	6	4.42366	4.05354	4.35224	4.13012	4.27950	4.20545	6	
	7	5.16094	4.72913	5.07762	4.81848	5.99275	4.90636	7	
	8	5.89821	5.40472	5.80299	5.50683	5.70600	5.60727	8	
	9	6.63549	6.08031	6.52836	6.19519	6.41925	6.30818	9	
45′	1	0.73432	0.67880	0.72236	0.69151	0.71018	0.70401	1	15′
	2	1.46864	1.35760	1.44472	1.38302	1.42037	1.40802	2	
	3	2.20296	2.03640	2.16709	2.07453	2.13055	2.11204	3	
	4	2.93729	2.71520	2.88945	2.76605	2.84074	2.81605	4	
	5	3.67161	3.39400	3.61182	3.45756	3.55092	3.52007	5	
	6	4.40593	4.07280	4.33418	4.14907	4.26111	4.22408	6	
	7	5.14025	4.75160	5.05654	4.84059	4.97129	4.92810	7	
	8	5.87458	5.43040	5.77891	5.53210	5.68148	5.63211	8	
	9	6.60890	6.10920	6.50127	6.22361	6.39166	6.33613	9	
Minutes	Distance	47° Dep.	47° Lat.	46° Dep.	46° Lat.	45° Dep.	45° Lat	Distance	Minutes

Differences of Latitude and Departure—Continued.

		45°			
	Lat.		Dep.		
1	0.70710		0.70710		1
2	1.41421		1.41421		2
3	2.12132		2.12132		3
4	2.82842		2.82842		4
5	3.53553		3.53553		5
6	4.24264		4.24264		6
7	4.94974		4.94974		7
8	5.65685		5.65685		8
9	6.36396		6.36396		9
	Dep.		Lat.		
		45°			

Chains, Yards, and Feet,

WITH THEIR RECIPROCAL EQUIVALENTS.

Link = 7.92 inches.   Chain = 66 feet = 792 inches.

CHAINS INTO FEET.				FEET INTO CHAINS.		
Chains.	Links.	Yards.	Feet.	Feet.	Yards.	Links.
0	1	0.22	0.66	0.10	.033	0.15
0	2	0.44	1.32	0.20	.066	0.30
0	3	0.66	1.98	0.25	.082	0.38
0	4	0.88	2.64	0.30	.010	0.45
0	5	1.10	3.30	0.40	.133	0.60
0	6	1.32	3.96	0.50	.166	0.76
0	7	1.54	4.62	0.60	.200	0.91
0	8	1.76	5.28	0.70	.233	1.06
0	9	1.98	5.94	0.75	.250	1.13
0	10	2.20	6.60	0.80	.266	1.21

Chains, Yards, and Feet—Continued.

Chains.	Links.	Yards.	Feet.	Feet.	Yards.	Links.
	CHAINS INTO FEET.			FEET INTO CHAINS.		
0	. 20	4.40	13.20	0.90	.300	1.36
0	. 30	6.60	19.80	1.00	.330	1.51
0	. 40	8.80	26.40	2.0	.660	3.0
0	. 50	11.00	33.00	3.0	1.000	4.5
0	. 60	13.20	39.60	4.0	1.330	6.0
0	. 70	15.40	46.20	5.0	1.66	7.5
0	. 80	17.60	52.80	6.0	2.00	9.1
0	. 90	19.80	59.40	7.0	2.33	10.6
1	. 00	22.00	66.00	8.0	2.66	12.1
2	. 00	44.00	132	9.0	3.00	13.6
3		66.00	198	10.0	3.33	15.1
4		88.00	264	15.0	5.00	22.7
5		110	330	20	6.66	30.3
6		132	396	24	8.00	36.3
7		154	462	27	9.00	40.9
8		176	528	30	10.00	45.4
9		198	594	33	11.00	50.0
10		220	660	36	12.00	54.5
20		440	1320	39	13.00	59.1
30		660	1980	40	13.33	60.6
35		770	2310	42	14.00	63.3
40		880	2640	45	15.00	68.2
45		990	2970	48	16.00	72.7
50		1100	3300	50	16.66	75.7
55		1210	3630	51	17.00	77.3
60		1320	3960	54	18.00	81.8
65		1430	4290	57	19.00	86.3
70		1540	4620	60	20.00	90.9
75		1650	4950	63	21.00	95.4
80		1760	5280	66	22.00	100

The Army Ration.

Table showing the weight and bulk of 1000 rations.

One thousand rations of	Nett weight in pounds.	Gross weight in pounds.	Bulk in barrels.	100 rations consist of
Pork - -	750.	1218.75	3.75	75 lbs. or ⎱
Bacon - -	750.	903.19	4.90	75 lbs. ⎰
Flour - -	1125.	1234.06	5.74	112.5 lbs. or ⎱
Pilot bread -	750.	921.69	9.03	75 lbs. or
Do. - - -	1000.	1228.91	12.05	100 lbs. in the field ⎰
Beans - -	155.	177.32	0.71	8 quarts, or ⎱
Rice - - -	100.	114.50	0.46	10 lbs. ⎰
Coffee - -	60.	70.90	0.35	6 lbs.
Sugar - -	120.	135.62	0.50	12 lbs.
Vinegar -	92.5	107.50	0.33	4 quarts.
Candles- -	15.	17.50	0.09	$1\frac{1}{2}$ lb.
Soap - -	40.	46.89	0.19	4 lbs.
Salt - - -	33.75	38.63	0.16	2 quarts.

Forage.

14 lbs. hay or fodder ⎱ per horse ⎱ when pressed 11 lbs. to cub. foot.
12 quarts oats, or   ⎰ per day. ⎰ 40 lbs. to bus., 33.14 lbs. cub. foot
8 quarts corn   55 lbs. to bus., 45.65 lbs. cub. foot

Daily allowance of water for a horse, 4 gallons.
Average mule pack, New Mexico, 175 lbs.
Average load to mule team across the Prairies, 2000 lbs.

TABLES AND FORMULÆ.

PART II.

GEODESY.

I. *Reduction to centre of station.*

Call P the place of the instrument,

    C the centre of the station,

    O the angle at P, between two objects A and B,

    y the angle at P, between C and the *left* hand object B,

    r the distance C P,

    C the unknown angle at C,

    D the distance A C,

    G the distance B C,

$$C = O + \frac{r \sin (O + y)}{D \sin 1''} - \frac{r \sin y}{G \sin 1''}$$

In the use of this formula proper attention should be paid to the signs of $\sin (O + y)$ and $\sin y$; for the first term will be *positive* when $(O + y)$ is less than 180°, (the reverse with $\sin y$); D being the distance of the *right* hand object, the graduation of the instrument running from left to right.

r being small, the lengths of D and G are computed with the angle O.

II. *Reduction to centre of signal observed, or correction for phase in tin cones used as signals.*

$$\text{Correction} = \pm \frac{r \cos^2 \tfrac{1}{2} Z}{D \sin 1''}$$

Where r = radius of the signal

    Z = angle at the point of observation between the Sun and the signal,

    D = the distance.

III. *Spherical Excess.*

$$E = \frac{S}{r^2 \sin 1''} = \frac{a\,b\,\sin C}{2\,r^2 \sin 1''}$$

S, being the area of the triangle. $r =$ the radius of the Earth

$$S = \frac{a\,b\,\sin c}{2}$$

$$= \sqrt{s\,(s-a)\,(s-b)\,(s-c)}, \ s \text{ being} = \frac{a+b+c}{2}$$

Between latitudes 45° and 25° the spherical excess amounts to about $1''$ for an area of 75.5 square miles.

Hence, if the area in square miles be known, a close approximation to the spherical excess will be had by dividing the area by 75.5.

Log. mean radius of the Earth in yards $= 6.8427917$.

If the three angles of a triangle are assumed to have been equally well determined, the previous determination of the spherical excess is not necessary for the calculation of the sides, though it will be required for estimating the relative accuracy of the observations.   For the sides of a spherical triangle may be computed as if they were rectilineal, when $\frac{1}{3}$ the excess of the sum of the three angles above 180° is deducted from each of the three observed angles;

then side $b =$ side $a \sin (B - \frac{1}{3} E) \div \sin (A - \frac{1}{3} E)$.

IV. *To reduce the length of an inclined base to horizontal measure.*

Let B be the length of the base on the inclined plane,

    b that reduced to the horizontal plane,

    θ the inclination,

$$b = \text{B} \cos \theta$$

But as θ is generally a small angle and need not be known with extreme precision, it is better to compute the excess of B above b, and supposing θ to be given in minutes.

$$\text{B}-b= \text{B} \left(1-\cos \theta\right) = 2 \text{ B} \sin^2\frac{\theta}{2} = \tfrac{1}{2} \text{ B } \theta^2 \sin^2 1' = \frac{\sin^2 1'}{2}\theta^2 \text{ B},$$

$$\text{or B} - b = 0.00000004231 \; \theta^2 \text{ B}$$

or by logarithms,

$$\text{Log (B}-b) = \text{const. log } 2.626422 + 2 \log \theta + \log \text{B}$$

V. *To reduce a broken base to a straight line.*

Let a and b be the given sides, and C the contained angle, very nearly 180°.

make $\text{C} = 180° - \theta$, θ being small, and $\cos \theta = 1 - \tfrac{1}{2}\theta^2$

$$\text{then, sine } c = a + b - \frac{\sin^2 1'}{2} \cdot \frac{a \; b \; \theta^2}{a+b} \; ,$$

$$= a + b - 0.00000004231 \times \frac{a \; b \; \theta^2}{a+b}$$

θ being expressed in minutes.

$$\text{log. } 0.00000004231 = 2.6264222$$

VI. *To find the length,* B D $= x$, *of a portion of a straight line* A H, *knowing the two other portions* A B $= a$; D H $= b$; *and also the angles* a, β, γ, *from any exterior station* C, *between* B *and* A, D *and* A, *and* H *and* A.

The problem being intended to supply by observation any portion of a base which cannot be directly measured.

$$\operatorname{tang}^2 \phi = \frac{4\,a\,b}{(a-b)^2} \; \frac{\sin \beta \sin (\gamma - a)}{\sin a \sin (\gamma - \beta)}$$

$$x = -\frac{a+b}{2} \pm \frac{a-b}{2 \cos \phi}$$

VII. *To reduce a measured base to the level of the sea.*

Let r represent the radius of the Earth (or better, the normal N,) corresponding to the base b at the level of the sea, and $r + a$ the radius referred to the level of the measured base B,

$$\text{then } r + a : r :: \text{B} : b = \text{B} \times \frac{r}{r + a}$$

$$\text{and B} - b = \text{B} - \text{B}\,\frac{r}{r+a} = \text{B} \times \left(\frac{a}{r} - \frac{a^2}{r^2} + \text{etc.} \right)$$

But the radius of the Earth being very great in comparison to the difference of level a, we have the correction δ sufficiently accurate by retaining only the first term. Hence,

$$\delta = \frac{\text{B}\,a}{r}$$

VIII. *Correction for temperature in metallic rods.*

Let e = the linear expansion for 1° of Fahrenheit,
l = the length of the rod before expansion,
l' = the length of the rod after expansion,
l = the number of degrees, Fahrenheit,

Total expansion = $e\,t$

and $\qquad l' = l\,(1 + e\,t)$

The following expansions were adopted by Mr. Hassler in his comparisons of weights and measures, (Report of 1832.)

Expansion for 1° Fahr. = e For 1° in a yard's length.

	Expansion for 1° Fahr. = e	For 1° in a yard's length	
Platinum	= 0.0000051344 ;	= 0.0001848384	Eng. In.
Brass Bar	= 0.00001050903;	= 0.00037832508	"
Iron Bar	= 0.000006963535;	= 0.000250687260	"

Other authorities :

	Expansion for 1° Fahr. = e	For 1° in a yard's length.		
Brass bar	0.000010480	0.0003772800	Eng. In.	Bailey.
Brass rod	0.0000105155	0.0003785580	"	Roy.
——	106666	0.0003839976	"	Troughton.
Brass wire	107407	0.0003866652	"	Smeaton.
Iron bar	0.0000069907	0.0002516652	"	Smeaton.
Steel rod	63596	0.0002289456	"	Roy.
Glass, Barom. tubes	43119	0.0001552284	"	Roy.
White Norway pine	22685	0.0000816660	"	Kater.

IX. *Measurement of distances by sound.*

The *velocity* of *Sound,* in one second of time at 32°
Fahrenheit in dry air, is about 1090 English feet. For
any higher temperature, add 1 foot for every degree of
the Thermometer above 32°.

The measurement of distances by sound should always
be made, if possible, in calm, *dry* weather. In cases of
wind, the velocity per second must be corrected by the
quantity, f cos d; f being the force of the wind in feet
per second, and d the angle which its direction makes
with that of the sound.

Or, in general, in dry air,

$$v = 1090 \text{ feet} + (t° - 32°) \pm f \cos d.$$

Velocity and force of winds.

Velocity in miles per hour.	A wind, when it does not exceed the velocity opposite to it, may be denominated	Velocity per sec'nd.	Force on a square foot.
		feet.	lbs.
6.8	a gentle, pleasant wind..................	10	0.129
13.6	a brisk gale.............................	20	0.915
19.5	a very brisk gale.......................	30	2.059
34.1	a high wind............................	50	5.718
47.7	a very high wind.......................	70	11.207
54.5	a storm or tempest.....................	80	14.638
68.2	a great storm	100	22.872
81.8	a hurricane	120	32.926
102.3	a violent hurricane, that tears up trees, etc.	150	51.426

X. *For Reconnoissances.*

" Three point problem."

At a point P, from whence are to be seen three points A, C, B, forming a triangle, the elements (i. e. the angles and sides) of which are known, measure the angles A P C, and C P B ; then, required to determine the direction and distance of the point P from each object.

Make A C $= a$; B C $= b$; B C A $=$ C ; A P C $=$ P, and C P B $=$ P′; also, make R $= 360°$ — P — P′ — C ; $x =$ C A P ; $y =$ P B C.

Then will

$$\text{Cot } x = \text{cot R} \left(\frac{a \sin \text{P}′}{b \sin \text{P} \cos \text{R}} + 1 \right)$$

$$y = \text{R} - x$$

The use of these formulæ need not be embarrassing if care is taken in properly applying the sign of cos. and cot. R. When R is less than 90° both cos. and cot. are plus ; between 90 and 180° both are minus ; between 180° and 270° the cos. is minus and the cot. plus ; between 270 and 360°, cos. is plus and the cot. minus.

This problem is indeterminate when P falls upon the circumference of the circle passing through A, B, C. A case of this nature is of rare occurrence, however, in practice.

XI. *For computing the principal Geodetic quantities depending on the spheroidal figure of the earth, at any given latitude.*

Eccentricity of the Earth $= e =$

$$= \left(\frac{a^2 - b^2}{a^2}\right)^{\frac{1}{2}} = \left(1 - \frac{b^2}{a^2}\right)^{\frac{1}{2}} = 2\,E - E^2$$

Ellipticity $= E \qquad = \dfrac{a - b}{a} \qquad = 1 - \dfrac{b}{a}$

or, very nearly $e^2 = 2\,E\,; \qquad E = \dfrac{e^2}{2}$

Normal ending at minor axis (or radius of curvature of a section perpendicular to the meridian)

$$= N = \frac{a}{(1 - e^2 \sin^2 L)^{\frac{1}{2}}}$$

Normal ending at major axis $= N' = N\,(1 - e^2)$

$$= \frac{a\,(1 - e^2)}{(1 - e^2 \sin^2 L)^{\frac{1}{2}}}$$

Tangent ending at minor axis $= t = N \cot L$

Tangent ending at major axis $= T = N \tan L\,(1 - e^2)$

Radius of the parallel $\qquad = \rho = N \cos L$

Radius of curvature of the merid. $= R = \dfrac{N^3}{a^2}(1 - e^2)$

$$= \frac{a\,(1 - e^2)}{(1 - e^2 \sin^2 L)^{\frac{3}{2}}}$$

Radius of curvature of a section making an angle Z with the meridian, $= R^z = \dfrac{N\,R}{N^2 \cos^2 Z + R^2 \sin^2 Z}$

Radius of the earth $= r = a\left(1 - \dfrac{e^2\,(1 - e^2)\,\sin^2 L}{1 - e^2 \sin^2 L}\right)^{\frac{1}{2}}$

$a =$ Equatorial Radius,
$b =$ Polar Radius,
$L =$ the given Latitude.

XII. *Numerical values of some of the preceding qnantities, from a discussion by* BESSEL *in the "Astronomische Nachrichten, No. 438."*

$a =$ Eq. Rad. $= 3272077.14$ toises; log $= 6.5148235337$
$b =$ Polar Rad. $= 3261139.33$ toises; log $= 6.5133693539$

Ratio of the Toise to the Metre—law of France, Dec. 10, 1799.

$$T = \overset{M}{1.9490363} \qquad Log = 0.2898199300$$

whence in metres

$$a = \overset{M}{6377397.15}; \qquad Log = 6.8046434637$$
$$b = \overset{M}{6356078.96}; \qquad Log = 6.8031892839$$

Ratio of the axes, $\quad a : b : : 299.1528 : 298.1528$;
mean uncertainty ± 4.667 units.

Length of the Earth's quad. $= \overset{T}{5131179.81} = \overset{M}{10000855.76}$;
mean uncertainty $= \pm 498.23$ metres.

$$e = \text{Eccentricity} = \left(1 - \frac{b^2}{a^2}\right)^{\frac{1}{2}} = 0.0816967;$$
$$Log = 8.9122052271$$
$$E = \text{Ellipticity} = \tfrac{1}{2} e^2 \qquad Log = 7.5233789824$$

Length, in toises, of a meridional degree whose middle latitude is φ.

$$D\,m = \overset{T}{57013.109} - \overset{T}{286.337} \cos 2\,\varphi + \overset{T}{0.611} \cos 4\,\varphi \left.\right\}$$
$$+ \overset{T}{0.001} \cos 6\,\varphi \left.\right\}$$

Length of a degree of the parallel, in toises,
$$D\,p = 57156.285 \cos \varphi - 47.825 \cos 3\,\varphi + 0.060 \cos 5\,\varphi$$
or making $\sin \psi = e \sin \varphi$
$$Log\ D\,p = 4.7567009.0 + \log \cos \varphi - \log \cos \psi$$

XIII. *Ratio of the Metre to the English Yard.*

The value of the French metre in English imperial inches, in general use in this country and in Europe, is that derived from Kater's Experiments in 1818, viz: 39.37079 inches of Sir G. Shuckburg's scale at 62° Faht., the metre being at 32° Faht.

From the more recent and accurate comparisons of Mr. Baily in 1835, when engaged in constructing a new standard scale for the Royal Astronomical Society (Mem. R. A. S., vol. ix); 39.369678 inches is the value of the standard metre, in mean inches of the centre yard of the Astronomical Society's scale, each being reduced to its standard temperature, namely, the platina metre to 32° and the brass scale to 62° of Fahrenheit's Thermometer. This very change of temperature, however, involves the result in some degree of uncertainty.

The centre yard of the Astronomical Society's scale exceeds the imperial standard yard by 0.000377 inches.

Whence, according to these experiments 39.370092 inches is the value of the standard metre in imperial standard inches, both being at their respective standard temperatures.

The value of the metre, as reported to Congress by Mr. Hassler in his report on Weights and Measures in 1832, is 39.38091714 inches of the English imperial standard at 32° Fht., the comparisons having been made at that temperature upon an 82 inches scale by Troughton, said to be *identical* with the English standard; or, correcting for expansion = 39.36850154 imperial standard inches at 62° Fht., a value materially smaller than the two preceding. According to Baily this discordance has probably arisen from inaccuracy in the length of the copy of Troughton's scale employed by Mr. Hassler.

This 82 inches scale is the standard of the United States, but in the absence of a direct comparison between it and the English standard, and not to add to a confusion already too great, it is as well to adhere for the present to the old value of Kater, as being that which is still most in use.   To recapitulate :

1 metre $=$ 39.3707900 English imperial inches, according to Kater, (1818,) Log $=$ 1.5951741293

$=$ 39.3700920 English imperial inches, according to Baily, (1835,) Log $=$ 1.5951664297

$=$ 39.36850154, American std'd inches, according to Hassler, (1832,) being the ratio, for the present, in use upon the Survey of the Coast, - - - - - Log $=$ 1.5951489169

The metre being at 32°, and the inches at 62° Fht.

XIV. *Numerical values of Bessel's terrestrial elements in English yards, adopting Kater's value of the metre,*
viz: 39.37079 English inches; Log 1.5951741293

Log to reduce toises to yards $=$ 0.3286915586
Log to reduce metres to yards $=$ 0.0388716286
Log . 3 $=$ 0.4771212547
Log . 12 $=$ 1.0791812460
Log 5280 $=$ 3.7226339225

$a=$ Equat. Rad. $=$ 6 974 532.339; Log $=$ 6.8435150923

$b=$ Polar Rad. $=$ 6 951 218.059 ; Log $=$ 6.8420609125

Length, in yards, of a Meridional degree, whose middle latitude is φ.

$$\mathrm{D}\,m = 121525.183\overset{\text{Y}}{.}\!183 - 610.336\overset{\text{Y}}{} \cos 2\,\phi + 1.302\overset{\text{Y}}{} \cos 4\,\phi$$
$$+ 0.002\overset{\text{Y}}{} \cos 6\,\phi$$

Length, in yards, of a degree of the parallel.

$$\mathrm{D}\,p = 121830.366\overset{\text{Y}}{} \cos \phi - 101.941\overset{\text{Y}}{} \cos 3\,\phi + 0.128\overset{\text{Y}}{} \cos 5\,\phi$$

or, making sin ψ = e sin φ

$$\mathrm{Log}\,\mathrm{D}\,p = 5.0853925 + \log \cos \phi - \log \cos \psi$$

or, using the logarithms of the numerical co-efficients,

$$\mathrm{D}\,m = 121525.183\overset{\text{Y}}{} - (2.7855691) \cos 2\,\phi$$
$$+ (0.1147) \cos 4\,\phi + (7.3287) \cos 6\,\phi$$

$$\mathrm{D}\,p = (5.0857556) \cos \phi - (2.00835) \cos 3$$
$$+ (9.1069) \cos 5\,\phi$$

$$\text{or, } \mathrm{D}\,p = \frac{(5.0853925)\ \cos \phi}{\cos \psi}$$

XV. *Constant Logarithms.*

$e^2 = 0.00667435$ . . . . Log $= 7.8244104542$

$\tfrac{1}{2}\,e^2 = \mathrm{E} = \text{Ellipticity} = \dfrac{1}{299.66}$ $= 7.5233789824$

Sin $1''$ . . . . $= 4.6855748668$

$\tfrac{1}{2}\sin 1''$ . . . . $= 4.3845448711$

$\dfrac{3\,e^2}{2}\sin 1''$ . . . . $= 2.6860751039$

$(1 - e^2) = 0.99332565$ . . . . . $= 9.9970916404$

$a\,(1 - e^2)$ . . . . $= 6.8406067325$

$a\sin 1''$ . . . . $= 1.5290899591$

$a\sin 1''$ (arith. comp.) . . . . . $= 8.4709100409$

XVI. *For computing the Geodetic Latitudes, Longitudes, and Azimuths of points of a Triangulation.*

1. In terms of the sides of the Triangles.*

$$u'' = \frac{K}{N \sin 1''} = \frac{K \, (1 - e^2 \sin^2 L)^{\frac{1}{2}}}{a \sin 1''}$$

$$\left.\begin{aligned} L' = L &- (1 + e^2 \cos^2 L) \, u'' \cos Z \\ &- (1 + e^2 \cos^2 L) \, (u'' \sin Z)^2 \, \text{tang} \, L \times \tfrac{1}{2} \sin 1'' \end{aligned}\right\}$$

$$M' = M + \frac{u'' \sin Z}{\cos L'}$$

$$Z' = 180° + Z - \frac{u'' \sin Z}{\cos L'} \, \sin \tfrac{1}{2} \, (L + L') \text{ or}$$

$$Z' = 180° + Z - (u'' \sin Z \, \text{tang} \, L + u''^2 \sin Z \cos Z \, \tfrac{1}{2} \sin 1''.)$$

2. In terms of the co-ordinates of rectangular axes referred to one of the points of the triangulation, the latitude and longitude of which are known; y being the ordinate in the direction of the meridian, and x the ordinate perpendicular to it.

---

*This is an abridgement of the following formulæ of Puissant, page 335, vol. 1, 3d edit.

$$\left.\begin{aligned} L' - L = &-\frac{K \cos Z}{N \sin 1''} \cdot \frac{N}{R} - \tfrac{1}{2} \frac{K^2 \sin^2 Z}{N^2 \sin 1''} \, \text{tang} \, L \cdot \frac{N}{R} \\ &+ \tfrac{1}{6} \frac{K^3}{N^3} \frac{\sin^2 Z \cos Z}{\sin 1''} (1 + 3 \, \text{tang}^2 \, L) \cdot \frac{N}{R} \end{aligned}\right\}$$

And $\dfrac{N}{R} = 1 + e^2 \cos^2 L + e^4 \cos^2 L + \tfrac{3}{2} e^2 \dfrac{K}{N} \cos Z \sin L \cos L$

$$L' = L \pm \frac{y}{R \sin 1''} - \tfrac{1}{2} \sin 1'' \left(\frac{x}{N \sin 1''} \right)^2 \times \left. \right\}$$

$$\text{tang} \left(L \pm \frac{y}{R \sin 1''} \right)$$

$$M' = M \pm \left(\frac{x}{N \sin 1''} \right) \times \frac{1}{\cos L'}$$

$$Z' = 270° \pm \frac{x}{N \sin 1''} \, \text{tang } L'$$

$K =$ distance in yards between two stations, the latitude and longitude of one of which is known, and w'' this same distance converted to seconds of arc.

$L =$ latitude of 1st station.

$M =$ longitude of 1st station, $+$ if west.

$Z =$ azimuth of 2d station at 1st, counted from the south round by the west, from 0° to 360°. The algebraic signs of the sine and cosine of this angle must be carefully attended to.

L', M', Z', the same things at 2d station, or quantities required.

$a =$ the equatorial radius; $e =$ the eccentricity.

$R =$ the radius of curvature of the meridian.

$N =$ the radius of curvature of a section perpendicular to the meridian.

The quantity $\dfrac{w'' \sin Z}{\cos L'} \sin \tfrac{1}{2} (L + L')$,

or $(M' - M) \sin \tfrac{1}{2} (L + L')$, by which the azimuth at one end of a line exceeds the azimuth at the other, is called *the convergence of the meridians.*

XVII. *To compute the length and direction of a line joining two points, the latitudes and longitudes of which are known, or measurement of a base by astronomical observations.*

$$\frac{\beta}{2} = \frac{e^2 \ (L - L') \ \cos^2 \tfrac{1}{2} \ (L + L')}{2}$$

$$N = \frac{a}{\left\{ \ 1 - e^2 \ \sin^2 \tfrac{1}{2} \ (L + L') \ \right\}^{\frac{1}{4}}}$$

$$l = L - \frac{\beta}{2} \qquad\qquad x'' = (M' - M) \ \cos l'$$

$$l' = L' + \frac{\beta}{2} \qquad\qquad y'' = (l - l') - \tfrac{1}{2} \sin 1'' \ x''^2 \ \text{tang } l$$

$$\text{tang } Z = \frac{x''}{y''} \qquad\qquad x = x'' \ N \ \sin 1''$$

$$u'' = \frac{x''}{\sin Z} = \frac{y''}{\cos Z} \qquad\qquad y = y'' \ N \ \sin 1''$$

$$K = u'' \ N \ \sin 1''$$

In which L, L', M, M', represent the latitudes and longitudes of the two points.

u'' = the distance between these points in seconds of arc.

K = the distance between these points in linear units.

x'' = the number of seconds in the arc passing through the point of which L' is the latitude, and perpendicular to the meridian of the point of which L is the latitude.

y'' = the seconds in the portion of this meridian between L and the foot of this perpendicular.

x, y = the same quantities in linear units.

Z = the azimuth of the second point L', from the first L.

N = the normal at the middle latitude.

Particular attention must be paid to the sign $(L - L')$ for upon this depends the sign of $\frac{\beta}{2}$, and also to that of $(l - l')$ in the value of y'', so as to know whether the small quantity $(-\frac{1}{2} \sin 1'' \, x''^2 \tang l)$ is to be added to or subtracted from $(l - l')$.

The azimuth Z is counted from the south round by the west, from $0°$ to $360°$.

The azimuth Z', (if required,) is to be computed from Z, as on page 65.

XVIII. *To compute the distance between two points, knowing their latitudes and the azimuth of one from the other.*

$$\frac{\beta}{2} = \frac{e^2 \, (L - L') \cos^2 \tfrac{1}{2} (L + L')}{2}$$

$$N = \frac{a}{\{ 1 - e^2 \sin^2 \tfrac{1}{2} (L + L') \}^{\frac{1}{2}}}$$

$$l = L - \frac{\beta}{2} \qquad\qquad \tang \phi = \frac{\tang l}{\cos Z}$$

$$l' = L' + \frac{\beta}{2} \qquad\qquad \sin (\phi - u'') = \frac{\sin l'}{\sin l} \sin \phi$$

$$K = u'' \, N \sin 1''$$

See the note to the preceding formulæ. The algebraic sign of the azimuth Z will determine the sign of ϕ, and consequently whether the quantity u'' is to be added to or subtracted from ϕ.

XIX. *To compute the distance between two points, knowing the latitude of one, the azimuth from this to the other, and the difference of their longitudes.*

$$\text{tang } \phi = \sin \text{L tang Z} \qquad \text{tang } L'' = \frac{\text{tang L} \sin (\phi - m)}{\sin \phi}$$

$$\beta = e^2 \, (\text{L} - \text{L}'') \cos^2 \tfrac{1}{2} \, (\text{L} + \text{L}'') \, ; \qquad \text{L}' = \text{L}'' - \beta$$

$$l = \text{L} - \frac{\beta}{2} \qquad\qquad l' = \text{L}' + \frac{\beta}{2}$$

$$u'' = \frac{m \cos l'}{\sin \text{Z}} \qquad\qquad \text{K} = u'' \, \text{N} \sin 1''$$

$m =$ the difference of longitude. The azimuth Z is, as before, counted from the south round by the west; its algebraic sign will determine the sign of ϕ, and consequently whether it is to be increased or diminished by m.

The formulæ on page 67 can be presented in a different form, thus :

From the formulæ on page 65,

$$(\text{M}' - \text{M}) \cos \text{L}' = u'' \sin \text{Z}$$

and,

$$u'' \cos \text{Z} = \frac{(\text{L} - \text{L}') - \tfrac{1}{2} u''^2 \sin^2 \text{Z} \cos^2 \text{L}' \tan \text{L} \sin 1'' \, (1 + e^2 \cos^2 \text{L})}{1 + e^2 \cos^2 \text{L}}$$

Substituting, in this last, the value of $u'' \sin \text{Z}$, and dividing one by the other ;

$$\text{tang Z} = \frac{(\text{M}' - \text{M}) \cos \text{L}' \, (1 + e^2 \cos^2 \text{L})}{(\text{L} - \text{L}') - \tfrac{1}{2} (\text{M}' - \text{M})^2 \cos^2 \text{L}' \, \text{tang L} \sin 1'' \, (1 + e^2 \cos^2 \text{L})}$$

Then knowing Z ;

$$u'' = \frac{(\text{M}' - \text{M}) \cos \text{L}'}{\sin \text{Z}}$$

and, $\qquad\qquad \text{K} = u'' \, \text{N} \sin 1''$

N, being the normal for the mean latitude.

XX. *Forms for record*

Survey of

No. of Triangle.	Position.	Names of Stations.	No. of Obs.	Observed Angles.			Errors and their dis-tribu'ion.	Spherical Angles.	Spherical Excess.	Final plane Angles.		
				°	′	″	″	″	′″	°	′	″
	Sought	*Cedar Point*	18	66	34	04.80	—0.36	04.44	1.58	66	34	02.86
XIII	Right.	*Buck Hill*	18	64	08	37.78	—0.36	37.42	1.58	64	08	35.84
	(Known Side.) Left.	*Fort Flats*	18	49	17	23.24	—0.36	22.88	1.58	47	17	21.30
										180	00	00.00

Example of

Survey of

NAMES OF STATIONS.	LATITUDES. $L' = L - u''.(1 + e^2 \, \mathrm{Cos.}^2 L) \, \mathrm{Cos.}\ Z$ $- \tfrac{1}{2} \mathrm{Sin.}\ 1'' \, \mathrm{Sin.}^2 Z \, u''^2 (1 + e^2 \, \mathrm{Cos.}^2 L) \, \mathrm{tang.}\ L.$
Fort Flats	Latitude $L = 45°39'13''.89$

$$
\begin{array}{llll}
\text{Log. K (yards)} & = 4.7295212 & \tfrac{1}{2}\,\text{Sin. } 1'' & = 4.38454 \\
\text{Log.} \dfrac{1}{\text{N Sin. } 1''} & = 8.4701676 & 2 \text{ Log. Sin. } Z & = 9.09522 \\
\\
\text{Log. } u'' & = 3.1996888 & 2 \text{ Log. } u'' & = 6.39936 \\
\text{Log.}(1+e^2\,\text{Cos.}^2 L) & = 0.0014140 & \ldots\ldots\ldots\ldots & = 0.00141 \\
\text{Log. Cos. } Z \quad (-) & = 9.9711240 & \text{Log. tang. } L & = 0.00991 \\
\\
\text{Log. 1st term} & = 3.1722268 & \text{Log. 2d term} & = 9.89034 \\
\\
\text{1st term} \quad (+) & = +1486''.71 & \text{2d term} & = \quad 0''.77 \\
\text{2d term} \quad (-) & = \quad -0.77 & & \\
\\
\delta L & = 0°\ 24'\ 45''.94 & & \\
L & = 45\ 39\ 13.89 & L + L' & = 91°\ 43'\ 13''.72 \\
& & \dfrac{L + L'}{2} & = 45\ 51\ 36.86 \\
\end{array}
$$

Cedar Point — Latitude $L' = 46°03'59''.83$

and computation.

Calculations of Triangles of the first order.

Computing Letter.	Logarithms of their Sines.	Calculation of the Sides.	Sides in Yards.	Designation.
S	9.9626198	Log. RL $\quad=4.7379524$ Comp. Log. Sin. S $=0.0373802$ Log. Sin. R $\quad=9.9541886$	$=54695.61$	{ *Buck Hill— Fort Flat.*
R	9.9541886	Log. LS $\quad=4.7295212$	$=53644.00$	{ *Fort Flat— Cedar Point.*
		Log. RL + Comp.Log.Sin.S $\}=4.7753326$ Log. Sin. L $\quad=9.8796760$		
L	9.8796760	Log. RS $\quad=4.6550086$	$=45186.49$	{ *Buck Hill— Cedar Point.*

Method 1, (page 65.)
Geodetic Determination of Positions. (Secondary.)

LONGITUDES $M' = M + \dfrac{u'' \, Sin. Z}{Cos. L'}$	AZIMUTHS $Z' = 180° + Z - (\delta M) Sin. \dfrac{L+L'}{2}$	REMARKS.
Lon. M $=84°42' 22''.19$	Azim. Z $=159°20' 13''.62$	
Log.Sin.Z$=(+)9.5476117$	$180°$	
Log u'' $\qquad=3.1996888$	$180° + Z \quad= 339\ 20\ 13.62$ $20\ 39\ 46.38$	
$\qquad\qquad 2.7473005$ Log. Cos. L' $=9.8412474$	Log.Sin. $\dfrac{L+L'}{2}= 9.8559089$	
Log. δ M $\qquad= 2.9060529$	$\ldots\ldots\ldots(+) = 2.9060529$	
	Log. δ Z $\qquad= 2.7619618$	
	$-\ 578''.05$	
δ M $\qquad= 0°13' 25''.48$	δ Z $\qquad= 0°09' 38''.05$	
M $\qquad= 84\ 42\ 22.19$	$180° + Z \quad= 339\ 20\ 13.62$	
Lon. M'$=84°55'47''.67$	Azim. Z'$=339°10' 35''.57$	

Ellipticity $= \dfrac{1}{300}$, Equat. Rad. $= 6974532$ yds. Log $= 6.8435151$

Latitude.		Normal or Radius of Curvature of the perpendicular to the Meridian. $N = \dfrac{a}{(1 - e^2 \sin^2 L)^{\frac{1}{2}}}$			Log. $(1 + e^2 \cos^2 L)$	
°	′	Log. N.	Com. differ'ce for 10′	Log. $\dfrac{1}{N \sin 1''}$		Differ. for 10′
20	0	6.8436847	27.3	8.4707404	0.0025521	55
	15	6888	27.6	7363	5439	56
	30	6929	27.8	7322	5356	55
	45	6971	28.1	7280	5274	56
21	0	7013	28.4	7238	5191	57
	15	7056	28.7	7196	5106	57
	30	7099	29.0	7153	5021	58
	45	7142	29.3	7109	4934	58
22	0	7186	29.5	7066	4847	58
	15	7230	29.7	7021	4760	59
	30	7274	30.0	6977	4671	59
	45	7319	30.2	6932	4582	60
23	0	7365	30.5	6887	4492	61
	15	7410	30.7	6841	4401	62
	30	7457	31.0	6795	4309	62
	45	7503	31.2	6748	4217	62
24	0	7550	31.5	6701	4124	63
	15	7597	31.7	6654	4030	63
	30	7645	32.0	6607	3935	63
	45	7693	32.2	6559	3840	64
25	0	7741	32.5	6510	3744	64
	15	7790	32.7	6462	3648	65
	30	7839	32.9	6413	3550	65
	45	7888	33.2	6363	3452	66
26	0	7938	33.4	6313	3353	66
	15	7988	33.6	6263	3254	67
	30	8038	33.8	6213	3154	67
	45	8089	34.0	6162	3053	68
27	0	8140	34.3	6111	2951	68
	15	8192	34.5	6060	2849	69
	30	8243	34.7	6008	2746	69
	45	8295	34.9	5956	2643	69
28	0	8348	35.1	5904	2539	70
	15	8400	35.3	5851	2434	70
	30	8453	35.5	5798	2329	71
	45	6.8438507	35.7	8.4705745	0.0022223	71

Latitude.		Log N.	Com. differ. for 10'	Log $\dfrac{1}{N \sin 1''}$	Log $(1 + e^2 \cos^2 L)$	Differ. for 10'
°	'					
29	0	6.8438560	35.9	8.4705691	0.0022117	71
	15	8614	36.1	5637	2010	72
	30	8668	36.3	5583	1902	72
	45	8723	36.5	5529	1794	72
30	0	8777	36.7	5474	1686	73
	15	8832	36.9	5419	1576	73
	30	8888	37.1	5364	1466	73
	45	8943	37.2	5308	1356	74
31	0	8999	37.4	5252	1245	74
	15	9055	37.6	5196	1134	75
	30	9111	37.7	5140	1022	75
	45	9168	37.9	5084	0910	75
32	0	9225	38.1	5027	0797	75
	15	9282	38.2	4970	0684	76
	30	9339	38.4	4912	0570	77
	45	9397	38.5	4855	0455	77
33	0	9454	38.7	4797	0340	77
	15	9512	38.9	4737	0225	77
	30	9571	39.0	4681	.0020109	77
	45	9629	39.1	4622	.0019993	77
34	0	9688	39.2	4564	9877	77
	15	9747	39.4	4505	9760	78
	30	9806	39.5	4446	9643	78
	45	9865	39.7	4387	9525	79
35	0	9924	39.8	4327	9407	79
	15	.8439984	40.0	4267	9288	80
	30	.8440044	40.0	4208	9169	80
	45	0104	40.1	4148	9050	80
36	0	0164	40.3	4087	8931	80
	15	0224	40.3	4027	8811	80
	30	0285	40.5	3966	8690	81
	45	0346	40.6	3906	8570	80
37	0	0406	40.7	3845	8449	81
	15	0467	40.7	3784	8328	81
	30	0529	40.9	3723	8206	81
	45	0590	41.0	3661	8084	81
38	0	0651	41.1	3600	7963	81
	15	0713	41.1	3538	7840	82
	30	0775	41.2	3477	7717	82
	45	0837	41.4	3415	7594	82
39	0	0898	41.4	3353	7471	82
	15	0961	41.4	3291	7348	82
	30	1023	41.5	3229	7224	83
	45	6.8441085	41.6	8.4703166	0.0017101	82
						83

Latitude		Log N.	Com. diff. for 10'	Log $\frac{1}{N \sin 1''}$	Log $(1+e^2 \cos^2 L)$	Differ. for 10'
°	′					
40	0	6.8441147	41.7	8.4703104	0.0016977	83
	15	1210	41.8	3041	6853	84
	30	1273	41.8	2979	6728	3
	45	1335	41.9	2916	6604	84
	0	1398	41.9	2853	6479	84
	15	1461	41.9	2791	6354	84
	30	1524	42.0	2728	6229	84
	45	1587	42.1	2665	6104	84
42	0	1650	42.1	2602	5979	84
	15	1713	42.1	2539	5853	84
	30	1776	42.2	2475	5728	84
	45	1839	42.1	2412	5602	84
3	0	1903	42.2	2349	5477	
	15	1967	42.2	2286	5351	
	30	2029	42.3	2222	5223	
	45	2093	42.3	2159	5099	
4	0	2156	42.3	2095	4973	
	15	2219	42.3	2032	4847	
	30	2283	42.3	1969	4721	
	45	2346	42.3	1905	4595	
45	0	2410	42.3	1842	4469	84
	15	2473	42.3	1778	4343	84
	30	2537	42.3	1715	4217	84
	45	2600	42.3	1651	4091	84
46	0	2663	42.3	1588	3965	84
	15	2727	42.3	1525	3839	84
	30	2790	42.3	1461	3713	84
	45	2854	42.2	1398	3587	84
47	0	2917	42.2	1334	3461	84
	15	2980	42.1	1271	3336	84
	30	3043	42.1	1208	3210	84
	45	3107	42.1	1145	3084	84
48	0	3170	42.1	1082	2959	84
	15	3233	42.0	1018	2833	84
	30	3296	42.0	0955	2708	84
	45	3359	41.9	0892	2583	84
49	0	3422	41.9	0830	2458	84
	15	3485	41.9	0767	2333	83
	30	3547	41.8	0704	2209	84
	45	3610	41.8	0641	2084	83
50	0	6.8443673		8.4700579	0.0011960	

Ellipticity $= \dfrac{1}{300}$, Equatorial Radius $= 6974532$ Yards.

Latitude.		Radius of curvature of Meridian. $R = \dfrac{a\,(1-e^2)}{(1-e^2 \sin^2 L)^{\frac{3}{2}}}$		
		Log R	Com. differ. for 10'.	Log $\dfrac{1}{R \sin 1''}$
°	′			
20	0	6.8411155		8.4733096
	15	1278	81.9	2973
	30	1402	82.7	2849
	45	1527	83.5	2724
21	0	1654	84.3	2598
	15	1781	85.1	2470
	30	1910	86.0	2341
	45	2040	86.8	2211
22	0	2172	87.5	2080
	15	2304	88.4	1947
	30	2438	89.1	1813
	45	2573	90.0	1679
23	0	2709	91.0	1543
	15	2846	91.5	1405
	30	2984	92.3	1267
	45	3124	93.0	1128
24	0	3264	93.6	0987
	15	3406	94.6	0845
	30	3549	95.2	0702
	45	3693	96.0	0559
25	0	3838	96.7	04¹4
	15	3984	97.4	026⸱
	30	4131	98.1	.4730120
	45	4279	98.8	.4729972
26	0	4428	99.4	9823
	15	4578	100.1	9673
	30	4730	100.9	9522
	45	4882	101.5	9370
27	0	5035	102.1	9216
	15	5189	102.8	9062
	30	5344	103.4	8907
	45	5500	104.1	8751
28	0	5657	104.7	8594
	15	5815	105.3	8436
	30	5974	106.0	8277
	45	6.8416134	106.5	8.4728117
			107.1	

XXI. *Trigonometrical Levelling*.

In the following formulæ let:

$\Delta \; \Delta'$ represent the observed zenith distance of which Δ is the smaller.

d A, d A$'$ the height of each signal above the telescope of the instrument.

K the distance in linear units between the two stations.

a the known altitude of the station from which the zenith distance Δ was measured.

N the normal for the mean latitude of the two stations.

M the modulus of common logarithms, having for its log 9.6377843.

1st. To compute the difference of level of two points by reciprocal zenith distances.

If possible, the zenith distances should be simultaneously taken, that the results may be independent of refraction.

$$\delta = \Delta + \frac{d \text{ A } \sin \Delta}{\text{K} \sin 1''} \qquad \delta = \Delta' + \frac{d \text{ A}' \sin \Delta'}{\text{K} \sin 1''}$$

$$\text{Log diff. of level} = \log \left\{ \text{K tang } \tfrac{1}{2}(\delta' - \delta) \right\} + \frac{\text{M}}{\text{N}} a \left. \right\}$$
$$\left. \pm \frac{\text{M}}{2\text{N}} \text{ K tang } \tfrac{1}{2}(\delta' - \delta) + \frac{\text{M}}{12 \text{ N}^2} \text{ K}^2 \right\}$$

The third term of this formula will be positive, if a is the altitude of the point from which the smallest zenith distance, *always represented by* Δ, has been observed; otherwise negative.

2d. To compute the difference of level of two points by a single zenith distance:

$$\delta = \Delta + \frac{d \, \text{A} \sin \Delta}{\text{K} \sin 1''}$$

$$\text{Log diff. of level} = \log \left\{ \frac{\text{K}}{\text{tang} \left(\delta - \frac{1-2\,r}{2\,\text{N} \sin 1''} \text{K} \right)} \right\}$$

$$+ \frac{\text{M}}{\text{N}} a \pm \frac{\text{M}}{2\,\text{N}} \left\{ \frac{\text{K}}{\text{tang} \left(\delta - \frac{1-2\,r}{2\,\text{N} \sin 1''} \text{K} \right)} \right\} + \frac{\text{M}}{12\,\text{N}^2} \text{K}^2$$

The third term will be positive when Δ is less than 90°, which will be the case when Δ is observed from the lowest point.

In this formula r represents the coefficient of terrestrial refraction, a variable quantity ; its mean value is generally stated to be 0.08 with variation of 0.02 less in summer and more in winter.

If we assume $r = 0.08$, the factor $\dfrac{1-2\,r}{2\,\text{N} \sin 1''} = 0''004133$;

assuming also N to be constant and $=$ to the normal at latitude 45°,

log N (in English feet) $= 7.3213623$;

$$\log \frac{\text{M}}{\text{N}} = 2.3164220 ; \qquad \log \frac{\text{M}}{2\,\text{N}} = 2.0153920 ;$$

$$\log \frac{\text{M}}{12\,\text{N}^2} = 3.9158785.$$

In ordinary cases, as an approximation, we may take:

difference of level $= \text{K} \cot (\Delta - 0.004133 \, \text{K.})$

K being in English feet and (0.004133 K) seconds of arc. $\qquad \text{Log } 0.004133 = 7.6163121.$

3d. The method of reciprocal zenith distances gives the means of obtaining the coefficient of refraction r, which, using the same notation as before, is :

$$r = \left\{ \frac{180° + \dfrac{K}{N \sin 1''} - (\delta + \delta')}{2\,\dfrac{K}{N \sin 1''}} \right.$$

In the trigonometrical survey of Massachusetts, Mr. Borden used 0.0784 as a mean coefficient for the sea coast, and 0.0697 for the interior of the State.

4th. To compute the altitude of a station from the observed zenith distance of the sea horizon ; using the same notation as before :

$$\log \text{Alt.} = \log \frac{N}{2}\left(\frac{\sin 1''}{1 - r}\right)^2 + \log\,(\delta - 90°)^2$$

$$+ \frac{M}{4}\left(\frac{\sin 1''}{1 - r}\right)^2 (\delta - 90°)^2$$

It would be as well, to ensure greater accuracy, to observe the zenith distance of points of the horizon on several days, taking a mean of the whole; and also to note the state of the tide at the time of observation.

Should r be assumed $= 0.08$

$$\log \tfrac{1}{2}\left(\frac{\sin 1''}{1 - r}\right)^2 = 9.1425441$$

$$\log \frac{M}{4}\left(\frac{\sin 1''}{1 - r}\right)^2 = 8.4792985$$

The last term can generally be neglected, and N may be assumed as the Normal of latitude 45°.

Corrections for Curvature and Refraction, showing the difference of the apparent and true level, in feet and decimals of a foot, for distances in feet and miles.

Distances in feet.	CORRECTION IN FEET.		
	For curvature.	For refraction.	For curvature and refract'n.
100	.00024	.00004	.00020
150	.00054	.00008	.00046
200	.00094	.00013	.00083
250	.00149	.00021	.00128
300	.00215	.00031	.00184
350	.00293	.00042	.00251
400	.00383	.00055	.00328
450	.00484	.00069	.00415
500	.00598	.00085	.00513
550	.00724	.00103	.00621
600	.00861	.00123	.00738
650	.01010	.00144	.00866
700	.01172	.00167	.01005
750	.01345	.00192	.01153
800	.01531	.00219	.01312
850	.01728	.00247	.01481
900	.01938	.00277	.01661
950	.02159	.00308	.01851
1000	.02392	.00333	.02059
1050	.02638	.00377	.02261
1100	.02895	.00414	.02481
1150	.03164	.00452	.02712
1200	.03445	.00492	.02953
1250	.03738	.00534	.03204
1300	.04043	.00578	.03465
1350	.04361	.00623	.03738
1400	.04689	.00670	.04019
1450	.05030	.00719	.04311
1500	.05383	.00769	.04614
1550	.05748	.00821	.04927
1600	.06125	.00875	.05250
1650	.06514	.00931	.05583
1700	.06914	.00988	.05926
1750	.07327	.01047	.06280
1800	.07792	.01107	.06645
1850	.08188	.01170	.07018
1900	.08637	.01234	.07403
1950	.09098	.01300	.07798
2000	.09570	.01367	.08203

Distances in miles.	CORRECTION IN FEET.		
	For curvature.	For refraction.	For curvature and refraction.
¼	.0417	.0060	.0357
½	.1668	.0238	.1430
¾	.3752	.0536	.3216
1	.6670	.0953	.5717
1½	1.5008	.2144	1.2864
2	2.6680	.3811	2.2869
2½	4.1688	.5955	3.5733
3	6.0030	.8561	5.1469
3½	8.1708	1.1673	7.0035
4	10.6720	1.5246	9.1474
4½	13.5468	1.9295	11.5773
5	16.6750	2.3821	14.2929
5½	20.1769	2.8824	17.2945
6	24.0120	3.4303	20.5817
6½	28.1809	4.0258	24.1551
7	32.6830	4.6690	28.0143
7½	37.5190	5.3599	32.1591
8	42.6880	6.0997	36.5883
8½	48.1910	6.8844	41.3066
9	54.0270	7.7181	46.3089
9½	60.1971	8.5996	51.5975
10	66.7000	9.5286	57.1714
11	80.7070	11.5296	69.1774
12	96.0480	13.7211	82.3269
13	112.7230	16.1033	96.6197
14	130.7320	18.6760	112.0560
15	150.0750	21.4393	128.6357
16	170.7520	24.3931	146.3589
17	192.7630	27.5376	165.2254
18	216.1086	30.8727	185.2359
19	240.7870	34.3981	206.3889
20	266.8000	38.1143	228.6857

For a very close approximation,

$$\text{correc'n for curvature, in ft.,} = \frac{2\,D^2}{3}$$

D being the distance in miles.

Reduction, in feet and decimals, upon 100 feet, for the following vertical angles.

Angle.	Reduct.	Angle.	Reduct.	Angle.	Reduct.	Angle.	Reduct.
° ′		° ′		° ′		° ′	
3 0	.137	7 30	.856	12 0	2.185	16 30	4.118
3 15	.161	7 45	.913	12 15	2.277	16 45	4.243
3 30	.187	8 0	.973	12 30	2.370	17 0	4.370
3 45	.214	8 15	1.035	12 45	2.466	17 15	4.498
4 0	.244	8 30	1.098	13 0	2.553	17 30	4.628
4 15	.275	8 45	1.164	13 15	2.662	17 45	4.760
4 30	.308	9 0	1.231	13 30	2.763	18 0	4.894
4 45	.343	9 15	1.300	13 45	2.866	18 15	5.030
5 0	.381	9 30	1.371	14 0	2.970	18 30	5.168
5 15	.420	9 45	1.444	14 15	2.077	18 45	5.307
5 30	.460	10 0	1.519	14 30	3.185	19 0	5.448
5 45	.503	10 15	1.596	14 45	3.295	19 15	5.591
6 0	.548	10 30	1.675	15 0	3.407	19 30	5.736
6 15	.594	10 45	1.755	15 15	3.521	19 45	5.882
6 30	.643	11 0	1.837	15 30	3.637	20 0	6.031
6 45	.663	11 15	1.921	15 45	3.754		
7 0	.745	11 30	2.008	16 0	3.874		
7 15	.800	11 45	2.095	16 15	3.995		

Ratio of Slopes for the following vertical angles.

Angle.	To one perpendicular.	Angle.	To one perpendicular.	Angle.	To one perpendicular.	Angle.	To one perpendicular.
° ′		° ′		° ′		° ′	
0 15	229	3 35	16	8 8	7	18 26	3
0 30	115	3 49	15	8 45	6¼	19 59	2¾
0 45	76	4 6	14	9 27	6	21 48	2½
1 0	57	4 24	13	9 52	5¾	23 58	2¼
1 15	46	4 45	12	10 18	5½	26 34	2
1 30	39	5 0	11½	10 47	5¼	29 44	1¾
1 45	33	5 12	11	11 19	5	33 42	1½
2 0	28	5 27	10½	11 53	4¾	38 40	1¼
2 15	25	5 42	10	12 32	4½	45 0	1
2 30	23	6 0	9½	13 15	4¼	53 8	¾
2 45	21	6 21	9	14 2	4	63 28	½
3 0	19	6 43	8½	14 55	3¾	75 58	¼
3 15	18	7 7	8	15 56	3½	78 41	⅕
3 28	17	7 36	7½	17 6	3¼		

XXII. *Barometrical Measurement of Heights.*

For computing the difference in the heights of two places, by means of the Barometer.

$$x = 60345.51 \left\{ 1 + . 001111 \left(t + t' - 64° \right) \right\}$$

$$\times \log \text{ of } \left\{ \frac{\beta}{\beta'} \times \frac{1}{1 + . 0001 \left(\tau - \tau' \right)} \right\}$$

$$\times \left\{ 1 + . 002695 \cos 2 \phi \right\}$$

Where ϕ = the latitude of the place.

β = the height of the barometer,
τ = the temperature (Faht.) of the mercury, ... } at the lower station.
t = the temperature (Faht.) of the air, ...

β' = the height of the barometer,
τ' = the temperature (Faht.) of the mercury, ... } at the upper station.
t' = the temperature (Faht.) of the air, ...

Make A = the log of the first term, in English feet.

B = the log of $1 + . 0001 \left(\tau - \tau' \right)$

C = the log of the last term.

D = $\log \beta - \left(\log \beta' + B \right)$

Then, by the tables which follow, the logarithm of the difference of altitude in English feet,

$$= A + C + \log D$$

TABLE I.—*Thermometers in the open air.*

$t+t'$	A.	$t+t'$	A.	$t+t'$	A.	$t+t'$	A.
°		°		°		°	
1	4.74914	46	4.77187	91	4.79348	136	4.81407
2	.74966	47	.77236	92	.79395	137	.81452
3	.75017	48	.77285	93	.79442	138	.81496
4	.75069	49	.77334	94	.79488	139	.81541
5	.75120	50	.77383	95	.79535	140	.81585
6	.75172	51	.77432	96	.79582	141	.81630
7	.75223	52	.77481	97	.79629	142	.81675
8	.75274	53	.77530	98	.79675	143	.81719
9	.75326	54	.77579	99	.79722	144	.81763
10	.75377	55	.77628	100	.79768	145	.81807
11	.75428	56	.77677	101	.79814	146	.81851
12	.75479	57	.77726	102	.79860	147	.81895
13	.75531	58	.77774	103	.79907	148	.81939
14	.75582	59	.77823	104	.79953	149	.81983
15	.75633	60	.77871	105	.79999	150	.82027
16	.75684	61	.77919	106	.80045	151	.82071
17	.75735	62	.77968	107	.80091	152	.82115
18	.75786	63	.78016	108	.80137	153	.82159
19	.75837	64	.78065	109	.80183	154	.82203
20	.75888	65	.78113	110	.80229	155	.82247
21	.75938	66	.78161	111	.80275	156	.82291
22	.75989	67	.78209	112	.80321	157	.82335
23	.76039	68	.78257	113	.80367	158	.82379
24	.76090	69	.78305	114	.80412	159	.82423
25	.76140	70	.78352	115	.80458	160	.82466
26	.76190	71	.78400	116	.80504	161	.82510
27	.76241	72	.78449	117	.80550	162	.82553
28	.76291	73	.78497	118	.80595	163	.82596
29	.76342	74	.78544	119	.80641	164	.82640
30	.76392	75	.78592	120	.80687	165	.82683
31	.76442	76	.78640	121	.80732	166	.82727
32	.76492	77	.78688	122	.80777	167	.82770
33	.76542	78	.78735	123	.80822	168	.82813
34	.76592	79	.78783	124	.80867	169	.82857
35	.76642	80	.78830	125	.80912	170	.82900
36	.76692	81	.78878	126	.80957	171	.82943
37	.76742	82	.78925	127	.81002	172	.82986
38	.76792	83	.78972	128	.81047	173	.83030
39	.76842	84	.79019	129	.81092	174	.83073
40	.76891	85	.79066	130	.81137	175	.83116
41	.76941	86	.79113	131	.81182	176	.83159
42	.76990	87	.79160	132	.81227	177	.83201
43	.77039	88	.79207	133	.81272	178	.83244
44	.77089	89	.79254	134	.81317	179	.83287
45	4.77138	90	4.79301	135	4.81362	180	4.83329

TABLE II.—*Attached Thermometer.*						TABLE III. *Latitude of the place.*	
$\tau-\tau'$	B.	$\tau-\tau'$	B.	$\tau-\tau'$	B.	φ	C.
°		°		°			
0	0.00000	20	0.00087	40	0.00174	0	0.00117
1	.00004	21	.00091	41	.00178	5	0.00115
2	.00009	22	.00096	42	.00182	10	0.00110
3	.00013	23	.00100	43	.00187	15	0.00100
4	.00017	24	.00104	44	.00191	20	0.00090
5	.00022	25	.00109	45	.00195	25	0.00075
6	.00026	26	.00113	46	.00200	30	0.00058
7	.00030	27	.00117	47	.00204	35	0.00040
8	.00035	28	.00122	48	.00208	40	0.00020
9	.00039	29	.00126	49	.00213	45	0.00000
10	.00043	30	.00130	50	.00217	50	9.99980
11	.00048	31	.00135	51	.00221	55	9.99960
12	.00052	32	.00139	52	.00226	60	9.99942
13	.00056	33	.00143	53	.00230	65	9.99925
14	.00061	34	.00148	54	.00234	70	0.99910
15	.00065	35	.00152	55	.00239	75	9.99900
16	.00069	36	.00156	56	.00243	80	9.99890
17	.00074	37	.00161	57	.00247	85	9.99885
18	.00078	38	.00165	58	.00252	90	9.99883
19	0.00083	39	0.00169	59	0.00256		

Example, latitude 21°.

	Upper Station.	Lower Station.
Thermometer in open air $t' = 70.\ 4$		$t = 77.\ 6$
Attached Thermometer.. $\tau' = 70.\ 4$		$\tau = 77.\ 6$
Barometer .... $\beta' = 23.66$		$\beta = 30.05$

$$B = 0.00031 \qquad \text{Log } D = 9.01502$$
$$\text{Log } \beta' = 1.37401 \qquad C = 0.00087$$
$$\text{———} \qquad A = 4.81939$$
$$1.37432 \qquad \text{———}$$
$$\text{Log } \beta = 1.47784 \qquad 3.83528$$
$$\text{———}$$
$$D = 0.10352 \qquad = 6843.7 \text{ feet.}$$

Table of comparison of Fahrenheit's Thermometer with Reaumur's and the Centesimal.

Fah.	Reaum.	Centes.	Fah.	Reaum.	Centes.	Fah.	Reaum.	Centes.
°	°	°	33	+ 0.4	+ 0.6	67	+15.6	+19.4
0	− 14.2	− 17.8	34	0.9	1.1	68	16.0	20.0
1	13.8	17.2	35	1.3	1.7	60	16.4	20.6
2	13.3	16.7	36	1.8	2.2	70	16.9	21.1
3	12.9	16.1	37	2.2	2.8	71	17.3	21.7
4	12.4	15.6	38	2.7	3.3	72	17.8	22.2
5	12.0	15.0	39	3.1	3.9	73	18.2	22.8
6	11.6	14.4	40	3.6	4.4	74	18.7	23.3
7	11.1	13.9	41	4.0	5.0	75	19.1	23.9
8	10.7	13.3	42	4.4	5.6	76	19.6	24.4
9	10.2	12.8	43	4.9	6.1	77	20.0	25.0
10	9.8	12.2	44	5.3	6.7	78	20.4	25.6
11	9.3	11.7	45	5.8	7.2	79	20.9	26.1
12	8.9	11.1	46	6.2	7.8	80	21.3	26.7
13	8.4	10.6	47	6.7	8.3	81	21.8	27.2
14	8.0	10.0	48	7.1	8.9	82	22.2	27.8
15	7.6	9.4	49	7.6	9.4	83	22.7	28.3
16	7.1	8.9	50	8.0	10.0	84	23.1	28.9
17	6.7	8.3	51	8.4	10.6	85	23.6	29.4
18	6.2	7.8	52	8.9	11.1	86	24.0	30.0
19	5.8	7.2	53	9.3	11.7	87	24.4	30.6
20	5.3	6.7	54	9.8	12.2	88	24.9	31.1
21	4.9	6.1	55	10.2	12.8	89	25.3	31.7
22	4.4	5.6	56	10.7	13.3	90	25.8	32.2
23	4.0	5.0	57	11.1	13.9	91	26.2	32.8
24	3.6	4.4	58	11.6	14.4	92	26.7	33.3
25	3.1	3.9	59	12.0	15.0	93	27.1	33.9
26	2.7	3.3	60	12.4	15.6	94	27.6	34.4
27	2.2	2.8	61	12.9	16.1	95	28.0	35.0
28	1.8	2.8	62	13.3	16.7	96	28.4	35.6
29	1.3	1.7	63	13.8	17.2	97	28.9	36.1
30	0.9	1.1	64	14.2	17.8	98	29.3	36.7
31	− 0.4	− 0.6	65	14.7	18.3	99	29.8	37.2
32	0.0	0.0	66	+15.1	+18.9	100	+30.2	+37.8

$x°$ Reaumur $= (32° + \tfrac{9}{4}x°)$ Fah. $= \tfrac{5}{4}x°$ Centes.

$x°$ Centes. $= (32° + \tfrac{9}{5}x°)$ Fah. $= \tfrac{4}{5}x°$ Reaum.

$x°$ Fah. $= (x° − 32°)\tfrac{4}{9}$ Reau. $= (x° − 32°)\tfrac{5}{9}$ Cen.

Table for the comparison of French and English Barometers.

Milli-metres.	English inches.	Millime-tres.	English inches.	Millime-tres.	English inches.
501	19.725	531	20.906	561	22.087
502	.764	532	.945	562	.126
503	.803	533	20.985	563	.166
504	.843	534	21.024	564	.205
505	.882	535	.063	565	.244
506	.921	536	.103	566	.284
507	19.961	537	.142	567	.323
508	20.000	538	.181	568	.363
509	.040	539	.221	569	.402
510	.079	540	.266	570	.441
511	.118	541	.300	571	.481
512	.158	542	.339	572	.520
513	.197	543	.378	573	.559
514	.236	544	.417	574	.599
515	.276	545	.457	575	.638
516	.315	546	.496	576	.678
517	.354	547	.536	577	.717
518	.394	548	.575	578	.756
519	.433	549	.614	579	.796
520	.473	550	.654	580	.835
521	.512	551	.693	581	.875
522	.551	552	.733	582	.914
523	.591	553	.772	583	.953
524	.630	554	.811	584	22.993
525	.670	555	.851	585	23.032
526	.709	556	.890	586	.071
527	.748	557	.930	587	.111
528	.788	558	21.969	588	.150
529	827	559	22.009	589	.189
530	20.867	560	22.048	590	23.229

Table for the comparison of French and English Barometers.

Millime-tres.	English inches.	Millime-tres.	English inches.	Millime-tres.	English inches.
591	23.268	621	24.449	651	25.630
592	.308	622	.489	652	.670
593	.347	623	.528	653	.709
594	.386	624	.567	654	.748
595	.426	625	.607	655	.788
596	.465	626	.646	656	.827
597	.504	627	.685	657	.867
598	.544	628	.725	658	.906
599	.583	629	.764	659	.945
600	.622	630	.804	660	25.985
601	.662	631	.843	661	26.024
602	.701	632	.882	662	.063
603	.741	633	.922	663	.103
604	.780	634	.961	664	.142
605	.819	635	25.000	665	.181
606	.859	636	.040	666	.221
607	.898	637	.079	667	.260
608	.937	638	.118	668	.300
609	23.977	639	.158	669	.339
610	24.016	640	.197	670	.378
611	.056	641	.237	671	.418
612	.095	642	.276	672	.457
613	.134	643	.315	673	.496
614	.174	644	.355	674	.536
615	.213	645	.394	675	.575
616	.252	646	.433	676	.615
617	.292	647	.473	677	.654
618	.331	648	.512	678	.693
619	.371	649	.552	679	.733
620	24.410	650	25.591	680	26.772

Table for the comparison of French and English Barometers.

Millime-tres.	English inches.	Millime-tres.	English inches.	Millime-tres.	English inches.
681	26.811	711	27.992	741	29.173
682	.851	712	28.032	742	.213
683	.890	713	.071	743	.252
684	.930	714	.110	744	.292
685	26.969	715	.150	745	.331
686	27.008	716	.189	746	.370
687	.048	717	.229	747	.410
688	.087	718	.268	748	.449
689	.126	719	.307	749	.488
690	.166	720	.347	750	.528
691	.205	721	.386	751	.567
692	.245	722	.425	752	.606
693	.284	723	.465	753	.646
694	.323	724	.504	754	.685
695	.363	725	.543	755	.725
696	.402	726	.583	756	.764
697	.441	727	.622	757	.803
698	.481	728	.662	758	.843
699	.520	729	.701	759	.882
700	.559	730	.740	760	.921
701	.599	731	.780	761	29.961
702	.638	732	.819	762	30.000
703	.677	733	.858	763	.040
704	.717	734	.898	764	.079
705	.756	735	.937	765	.118
706	.795	736	28.977	766	.158
707	.835	737	29.016	767	.197
708	.874	738	.055	768	.236
709	.914	739	.095	769	.276
710	27.953	740	29.134	770	30.315

12

*Table for the comparison of French and English
Barometers.*

Millimetres.	English inches.	Millimetres.	English inches.	PROPORTIONAL PARTS.	
				Millim.	English inches.
771	30.355	781	30.748	0.1	0.0039
772	.394	782	.788	.2	.0079
773	.433	783	.827	.3	.0118
774	.473	784	.866	.4	.0157
775	.512	785	.906	.5	.0197
776	.551	786	.945	.6	.0236
777	.591	787	30.984	.7	.0276
778	.630	788	31.024	.8	.0315
779	.670	789	.063	0.9	.0354
780	30.709	790	31.103	1.0	0.0394

I Metre　　　=　39.3707　English inches = 443.296　Paris lines.
I English foot =　0.304794 metre　　　= 135.114　Paris lines.
I French foot =　1.0658　English feet　=　0.32484 metre.

French inches	English inches.	French lines.	English inches.
1	1.0658	1	0.0888
2	2.1315	2	.1776
3	3.1973	3	.2664
4	4.2631	4	.3553
5	5.3288	5	.4441
6	6.3946	6	.5329
7	7.4604	7	.6217
8	8.5261	8	.7105
9	9.5919	9	.7993
10	10.6577	10	.8881
11	11.7234	11	.9770
12	12.7899	12	1.0658

Table for the comparison of French and English Barometers.

English inches and tenths.	Hundredths of an inch.				
	0	2	4	6	8
	Millimetres.				
21.0	533.39	533.90	534.41	534.91	535.42
.1	535.93	536.44	536.95	537.45	537.96
.2	538.47	538.98	539.49	539.99	540.50
.3	541.01	541.52	542.03	542.53	543.04
.4	543.55	544.06	544.57	545.07	545.58
.5	546.09	546.60	547.11	547.61	548.12
.6	548.63	549.14	549.65	550.15	550.66
.7	551.17	551.68	552.19	552.69	553.20
.8	553.71	554.22	554.73	555.23	555.74
.9	556.25	556.76	557.27	557.77	558.28
22.0	558.79	559.30	559.81	560.31	560.82
.1	561.33	561.84	562.35	562.85	563.36
.2	563.87	564.38	564.89	565.39	565.90
.3	566.41	566.92	567.43	567.93	568.44
.4	568.95	569.46	569.97	570.47	570.98
.5	571.49	572.00	572.51	573.01	573.55
.6	574.03	574.54	575.05	575.55	576.06
.7	576.57	577.08	577.59	578.09	578.60
.8	579.11	579.62	580.13	580.63	581.14
.9	581.65	582.16	582.67	583.17	583.68
23.0	584.19	584.70	585.21	585.71	586.22
.1	586.73	587.24	587.75	588.25	588.76
.2	589.27	589.78	590.29	590.79	591.30
.3	591.81	592.32	592.83	593.33	593.84
.4	594.35	594.86	595.37	595.87	596.38
.5	596.89	597.40	597.91	598.41	598.92
.6	599.43	599.94	600.45	600.95	601.46
.7	601.97	602.22	602.99	603.49	604.00
.8	604.51	605.02	605.53	606.03	606.54
.9	607.05	607.56	608.07	608.57	609.08

Table for the comparison of French and English Barometers.

English inches and tenths.	Hundredths of an inch.				
	0	2	4	6	8
	Millimetres.				
24.0	609.59	610.10	610.61	611.11	611.62
.1	612.13	612.64	613.15	613.65	614.16
.2	614.67	615.18	615.69	616.19	616.70
.3	617.21	617.72	618.23	618.73	619.24
.4	619.75	620.26	620.77	621.27	621.78
.5	622.29	622.80	623.31	623.81	624.32
.6	624.83	625.34	625.85	626.34	626.86
.7	627.37	627.88	628.39	628.89	629.40
.8	629.91	630.42	630.93	631.43	631.94
.9	632.45	632.96	633.47	633.97	634.48
25.0	634.99	635.50	636.01	636.51	637.02
.1	637.53	638.04	638.55	639.05	639.56
.2	640.07	640.58	641.09	641.59	642.10
.3	642.61	643.12	643.63	644.13	644.64
.4	645.15	645.66	646.17	646.67	647.18
.5	647.69	648.20	648.71	649.21	649.72
.6	650.23	650.74	651.25	651.75	652.26
.7	652.77	653.28	653.79	654.29	654.80
.8	655.31	655.82	656.33	656.83	657.34
.9	657.85	658.36	658.87	659.37	659.88
26.0	660.39	660.90	661.41	661.91	662.42
.1	662.93	663.44	663.95	664.45	664.96
.2	665.47	665.98	666.49	666.99	667.50
.3	668.01	668.52	669.03	669.53	670.04
.4	670.55	671.06	671.57	672.07	672.58
.5	673.09	.673.60	674.11	674.41	675.12
.6	675.63	676.14	676.65	677.15	677.66
.7	678.17	678.68	679.19	679.69	680.20
.8	680.71	681.22	681.73	682.23	682.74
.9	683.25	683.76	684.27	684.77	685.28

Table for the comparison of French and English Barometers.

English inches and tenths.	Hundredths of an inch.				
	0	2	4	6	8
	Millimetres.				
27.0	685.79	686.30	686.81	687.31	687.82
.1	688.33	688.84	689.35	689.85	690.36
.2	690.87	691.38	691.89	692.39	692.90
.3	693.41	693.92	694.43	694.93	695.44
.4	695.95	696.46	696.97	697.47	697.98
.5	698.49	699.00	699.51	700.01	700.52
.6	701.03	701.54	702.05	702.55	703.06
.7	703.57	704.08	704.59	705.09	705.60
.8	706.11	706.62	707.13	707.63	708.14
.9	708.65	709.16	709.67	710.17	710.68
28.0	711.19	711.70	712.21	712.71	713.22
.1	713.73	714.24	714.75	715.25	715.77
.2	716.27	716.78	717.29	717.79	718.30
.3	718.81	719.32	719.83	720.33	720.84
.4	721.35	721.86	722.37	722.87	723.38
.5	723.89	724.40	724.91	725.41	725.92
.6	726.43	726.94	727.45	727.95	728.46
.7	728.97	729.48	729.99	730.49	731.00
.8	731.51	732.02	732.53	733.03	733.54
.9	734.05	734.56	735.07	735.57	736.08
29.0	736.59	737.10	737.61	738.11	738.62
.1	739.13	739.64	740.15	740.65	740.16
.2	741.67	742.18	742.69	743.19	743.70
.3	744.21	744.72	745.23	745.73	746.24
.4	746.75	747.26	747.77	748.27	748.78
.5	749.29	749.80	750.31	750.81	751.32
.6	751.83	752.34	752.85	753.35	753.86
.7	754.37	754.88	755.39	755.89	756.40
.8	756.91	757.42	757.93	758.43	758.94
.9	759.45	759.96	760.47	760.97	761.48

Table for the comparison of French and English Barometers.

English inches and tenths.	Hundredths of an inch.				
	0	2	4	6	8
	Millimetres.				
30.0	761.99	762.50	763.01	763.51	764.02
.1	764.53	765.04	765.55	766.05	766.56
.2	767.07	767.58	768.09	768.59	769.10
.3	769.61	770.12	770.63	771.13	771.64
.4	772.15	772.66	773.17	773.67	774.18
.5	774.69	775.20	775.71	776.21	776.72
.6	777.23	777.74	778.25	778.75	779.26
.7	779.77	780.28	780.79	781.29	781.80
.8	782.31	782.82	783.33	783.83	784.34
.9	784.85	785.36	385.87	786.37	786.88

Table of corrections for capillary action to be added to English Barometers.

Diameter of tube.	Ivory.	Young.	Laplace.	Com. of Physics, &c. Royal Soc., 1840.	
				Unboiled tubes.	Boiled tubes.
Inches.	Inches.	Inches.	Inches.	Inches.	Inches.
0.05	0.2949	0.2964	0.		
.10	.1404	.1424	.1394	0.142	0.070
.15	.0865	.0880	.0854	.088	.044
.20	.0583	.0589	.0580	.060	.029
.25	.0409	.0404	.0412	.040	.020
.30	.0293	.0280	.0296	.028	.014
.35	.0212	.0196	.0216	.020	.010
.40	.0154	.0139	.0159	.014	.007
.45	.0112	.0100	.0117	.010	.005
.50	.0082	.0074	.0087	.007	.003
.60	.0043	.0645	.0046	0.004	0.002
.70	.0023	....	.0024		
0.80	0.0012	....	0.0013		

XXIII. *Thermometrical Measurement of Heights.*

Table of Barometric pressures corresponding to temperatures of boiling water.

Degrees of Fahrenheit.	TENTHS OF A DEGREE OF FAHRENHEIT.				
	0.	2.	4.	6.	8.
185	17.048	17.123	17.199	17.274	17.350
186	.425	.502	.578	.655	.731
187	.808	.886	.964	18.042	18.120
188	18.198	18.277	18.357	.436	.516
189	.595	.676	.756	.837	.918
190	.999	19.081	19.163	19.245	19.328
191	19.410	·493	.577	.661	.744
192	.828	.913	.998	20.083	20.169
193	20.254	20.341	20.427	.514	.601
194	.688	.776	.864	.952	21.041
195	21.129	21.219	21.309	21.398	.488
196	.578	.669	.761	.853	.944
197	22.036	22.129	22.222	22.315	22.409
198	.502	.597	.692	.786	.881
199	.976	23.072	23.169	23.265	23.362
200	23.458	.556	.654	.752	.850
201	.948	24.047	24.147	24.247	24.346
202	24.446	.547	.648	.750	851
203	.952	25.055	25.158	25.261	25.364
204	25.467	.572	.677	.781	.886
205	.991	26.097	26.204	26.311	26.417
206	26.524	.632	.741	.849	.957
207	27.066	27.176	27.286	27.397	27.507
208	.617	.729	.841	.954	28.066
209	28.178	28.292	28.406	28.521	.635
210	.749	.865	.981	29.098	29.214
211	29.330	29.448	29.566	.685	.803
212	.921	30.041	30.161	30.281	30.402

XXIV. *Formulæ for computing the elements for the projection of Maps.*

1. *For surfaces of not more than four degrees of latitude
and longitude, the formulæ being approximate.*

1. Normal $= N$ $= \dfrac{a}{(1 - e^2 \sin^2 L)^{\frac{1}{2}}}$

2. Tangent $= T$ $= N \cot L$

3. Radius of the parallel

         $= (R p)$ $= N \cos L$

4. Degree of the parallel

$$= (D p) \quad = \frac{(R p) \pi}{180°}$$

5. Number of minutes of the parallel

$$= (n') p \quad = (n') p \frac{D p}{60}$$

6. Angle between the tangent and the chord

$$= \sin \tfrac{1}{2} Z \quad = \frac{(n') p}{2 T}$$

Then,

difference of parallels $= y = \delta p = (n') p \sin \tfrac{1}{2} z$
$$= \frac{[(n') p]^2}{2 T}$$

difference of meridians $= x = \delta m = (n') p \cos \tfrac{1}{2} z$

The values δm and δp will be found in the following
tables.

Example of their use.—Let it be required to make a projection containing 40' of longitude between the parallels
of 41° 30' and 42° 10', to be subdivided to 5'.

Assume the centre of the sheet to be the intersection of
the middle parallel with the middle meridian of the pro-
posed map, which point call A; in this case a point in the
parallel of 41° 50'.

Through A draw the central meridian and a line at right angles to it.

Beginning at A, lay off above and below, on the central meridian, the values of D m from 41° 50′ to 41° 55′; 41° 55′ to 42°; 42° to 42° 5′, etc.; and from 41° 50′ to 41° 45′; 41° 45′ to 41° 40′, etc. These values to be taken from the table of *Meridional arcs—values of D m in yards*, by interpolation from the values there given for the middle latitudes of 41° and 42°.

Through each of the points ....A″.A′.A.A$_\iota$ A$_{\iota\iota}$... thus found, lay off perpendiculars to the central meridian.

Now turn to the table of *Co-ordinates δ m and δ p in yards* and lay off, from each of the points ....A″.A′.A.A$_\iota$ A$_{\iota\iota}$... to the right and left of the central meridian, the values of δ m for successively 5′, 10′, 15′ and 20′, corresponding (by interpolation from the columns of 41° 30′ and 42°) to each parallel of latitude required; and, from the points thus found, the corresponding values of δ p at right angles to the lines already drawn.

Lines passing through the extremities of δ p will be the required meridians and parallels.

The projection being made, any point whose latitude and longitude is known will be projected on the map from elements taken from the table of values of D m and D p, which are measured from the *meridians* and *parallels*, and not from the axes of co-ordinates used in making the projection.

2. *When the map extends over several degrees of latitude and longitude,* the preceding approximate formulæ will not answer; a middle latitude is assumed where the developing cone is tangent, and the projection made as follows:

Through the centre of the map, A, two lines are drawn, A y representing the central meridian, and the other, A x,

13

a line perpendicular to it; from the point A, along the line A y (above and below A) the lengths $s . s_1 s_2 s_3 \ldots$ $s r$ (in miles or yards) of degrees or minutes, as the case may be, of the meridian are laid down, and the remaining intersections of meridians and parallels are projected by means of co-ordinates y and x from the central point A, as follows:

$$x = \mathrm{d}\, m = \text{difference of meridians} = \rho \sin \theta$$

$$y = \mathrm{d}\, p = \text{difference of parallels} = s + x \tan \tfrac{1}{2} \theta$$

$$\theta = (n') p \frac{\mathrm{R}\, p}{\mathrm{T}} \qquad\qquad \rho = \mathrm{T} \pm \mathrm{S}$$

where $s =$ the length on the meridian from minute to minute or degree to degree as desired.

$\mathrm{T} = \mathrm{N} \cot \mathrm{L} =$ the tangent at the central point of the map, L being the latitude and N the normal at that point.

$(\mathrm{R}p) = \mathrm{N}' \cos l =$ the radius of the parallel at any point of latitude (above or below the central point) of which the ordinate is required.

$(n') p =$ the number of minutes of the parallel of the new point of which the ordinate is required.

3. *In maps of large portions of the Earth's surface* deviations from real magnitudes may be lessened by making the developing cone cut two parallels equidistant from the middle parallel; say through one-third of the length of the middle meridian of the map.

It will then be necessary to substitute for T in the preceding equation, the distance from the vertex of the cone to either intersection of the Earth's surface $= \dfrac{\mathrm{R}\, p}{\sin \mathrm{S}}$, S being the angle at the vertex between the elements of the cone and its axis; equal, in a spheroid, to one-half the sum of the geocentric latitudes of the two points of intersection.

Co-ordinates, δm, δp, in Yards.

Long. value of Z.	Lat. 22° 0′		Lat. 22° 30′		Lat. 23° 0′	
	δm	δp	δm	δp	δm	δp
1′	1882.0	0.1	1875.3	0.1	1868.5	0.1
2	3763.9	0.4	3750.6	0.4	3737.0	0.4
3	5645.9	0.9	5625.9	1.0	5605.4	1.0
4	7527.8	1.6	7501.2	1.7	7473.9	1.7
5	9409.9	2.6	9376.4	2.6	9342.4	2.7
6	11291.8	3.7	11251.7	3.7	11210.9	3.8
7	13173.7	5.0	13127.0	5.1	13079.4	5.2
8	15055.7	6.6	15002.3	6.7	14947.8	6.8
9	16937.6	8.3	16877.6	8.5	16816.3	8.6
10	18819.6	10.3	18752.9	10.5	18684.8	10.6
11	20701.6	12.4	20628.2	12.6	20553.3	12.8
12	22583.5	14.8	22503.5	15.0	22421.8	15.3
13	24465.5	17.3	24378.8	17.6	24290.2	17.9
14	26347.4	20.1	26254.1	20.5	26158.7	20.8
15	28229.4	23.1	28129.3	23.5	28027.2	23.9
16	30111.4	26.2	30004.6	26.7	29895.7	27.2
17	31993.3	29.6	31879.9	30.2	31764.2	30.7
18	33875.3	33.2	33755.2	33.3	33632.6	34.4
19	35757.2	37.0	35630.5	37.7	35501.1	38.3
20	37639.2	41.0	37505.8	41.7	37369.6	42.5
25	47049.0	64.1	46675.8	65.2	46712.0	66.4
30	56458.7	92.3	56258.7	93.9	56054.3	95.6
40	75278.2	164.1	75011.5	167.0	74730.0	169.9
50	94097.7	256.3	93764.2	260.9	93423.7	265.5
1° 00	112917.0	369.1	112516.9	375.8	112108.2	382.3
1 20	150555.4	656.2	150021.9	668.0	149476.9	679.6
1 30	169374.4	830.5	168774.2	845.4	168161.1	860.1
1 40	188193.3	1025.3	187526.3	1043.8	186845.1	1061.8
2 00	225830.5	1476.5	225030.0	1503.0	224212.5	1529.1
2 30	282284.7	2307.1	281284.0	2348.5	280261.9	2389.2
3 00	338736.6	3322.2	337535.6	3381.8	336309.0	3440.4
3 30	395186.0	4521.9	393784.5	4603.0	392353.1	4682.8
4 00	451632.0	5906.2	450029.9	6012.1	448393.7	6116.2

Co-ordinates, δm, δp, in Yards.

Long. value of Z.	Lat. 23° 30′		Lat. 24° 0′		Lat. 24° 30′	
	δm	δp	δm	δp	δm	δp
1′	1861.5	0.1	1854.4	0.1	1847.2	0.1
2	3723.1	0.4	3708.9	0.4	3694.4	0.4
3	5584.6	1.0	5563.3	1.0	5541.6	1.0
4	7446.1	1.8	7417.7	1.8	7388.8	1.8
5	9307.6	2.7	9272.2	2.7	9236.0	2.8
6	11169.2	3.9	11126.6	3.9	11083.2	4.0
7	13030.7	5.3	12981.0	5.4	12930.4	5.4
8	14892.2	6.9	14835.5	7.0	14777.6	7.1
9	16753.7	8.7	16689.9	8.9	16624.8	9.0
10	18615.3	10.8	18544.3	11.0	18472.0	11.1
11	20476.8	13.1	20398.8	13.3	20319.2	13.5
12	22338.3	15.6	22253.2	15.8	22166.4	16.0
13	24199.9	18.2	24107.6	18.5	24013.6	18.8
14	26061.4	21.2	25962.1	21.5	25860.8	21.8
15	27922.9	24.3	27816.5	24.7	27708.0	25.1
16	29784.4	27.7	29670.9	28.1	29555.2	28.5
17	31646.0	31.2	31525.4	31.7	31402.4	32.2
18	33507.5	35.0	33379.8	35.5	33249.6	36.0
19	35369.0	39.0	35234.2	39.6	35096.8	40.2
20	37230.5	43.2	37088.7	43.9	36944.0	44.6
25	46538.1	67.5	46360.8	68.6	46179.9	69.6
30	55845.8	97.2	55632.9	98.7	55415.9	100.3
40	74460.9	172.7	74177.1	175.5	73887.7	178.3
50	93076.0	269.9	92721.3	274.3	92359.5	278.5
1° 00	111691.0	388.7	111265.3	394.9	110831.4	401.1
1 20	148920.6	690.9	148353.0	702.1	147774.1	713.0
1 30	167535.2	874.5	166896.6	888.6	166245.4	902.4
1 40	186149.7	1079.6	185440.1	1097.0	184716.5	1114.1
2 00	223377.9	1554.6	222526.4	1579.7	221658.0	1604.3
2 30	279218.6	2429.1	278154.1	2468.3	277068.4	2506.8
3 00	335056.8	3497.9	333779.1	3554.4	332476.1	3609.8
3 30	390892.0	4761.1	389401.1	4837.9	387880.6	4913.3
4 00	446723.4	6218.5	445019.2	6318.9	443281.1	6417.4

Co-ordinates, δ m, δ p, in Yards.

Long. value of Z.	Lat. 25° 0′		Lat. 25° 30′		Lat. 26° 0′	
	δ m	δ p	δ m	δ p	δ m	δ p
1′	1839.8	0.1	1832.3	0.1	1824.7	0.1
2	3679.6	0.5	3664.6	0.5	3649.3	0.5
3	5519.5	1.0	5496.9	1.0	5474.0	1.0
4	7359.3	1.8	7329.2	1.8	7298.6	1.9
5	9199.1	2.8	9161.5	2.9	9123.3	2.9
6	11038.9	4.1	10993.8	4.1	10947.9	4.2
7	12878.8	5.5	12826.1	5.6	12772.6	5.7
8	14718.6	7.2	14658.5	7.3	14597.2	7.4
9	16558.4	9.2	16490.8	9.3	16421.9	9.4
10	18398.2	11.3	18323.1	11.5	18246.5	11.6
11	20238.0	13.7	20155.4	13.9	20071.2	14.1
12	22077.9	16.3	21987.7	16.5	21895.8	16.8
13	23917.7	19.1	23820.0	19.4	23720.5	19.7
14	25757.5	22.2	25652.3	22.5	25545.1	22.8
15	27597.3	25.4	27484.6	25.8	27369.8	26.2
16	29437.1	29.0	29316.9	29.4	29194.4	29.8
17	31277.0	32.7	31149.2	33.2	31019.1	33.6
18	33116.8	36.6	32981.5	37.2	32843.7	37.7
19	34956.6	40.8	34813.8	41.4	34668.4	42.0
20	36796.4	45.2	36646.1	45.9	36493.0	46.5
25	45995.5	70.7	45807.6	71.7	45616.2	72.7
30	55194.6	101.8	54969.1	103.2	54739.5	104.7
40	73592.7	180.9	73292.0	183.6	72985.8	186.1
50	91990.7	282.7	91614.9	286.8	91232.1	290.8
1° 00	110388.6	407.1	109937.6	413.0	109478.3	418.8
1 20	147184.0	723.8	146582.7	734.3	145970.3	744.6
1 30	165581.5	916.0	164905.0	929.3	164216.0	942.3
1 40	183978.8	1130.9	183227.1	1147.3	182461.5	1163.4
2 00	220772.7	1628.5	219870.6	1652.1	218951.9	1675.2
2 30	275961.6	2544.5	274833.9	2581.4	273685.3	2617.6
3 00	331147.8	3664.1	329794.3	3717.3	328415.8	3769.3
3 30	386330.6	4987.2	384751.3	5059.6	383142.7	5130.5
4 00	441509.4	6513.9	439704.0	6608.5	437865.3	6701.0

Co-ordinates, δm, δp, in Yards.

Long. value of Z	Lat. 26° 30'		Lat. 27° 0'		Lat. 27° 30'	
	δm	δp	δm	δp	δm	δp
1'	1816.9	0.1	1808.9	0.1	1800.9	0.1
2	3633.7	0.5	3617.9	0.5	3601.7	0.5
3	5450.6	1.1	5426.8	1.1	5402.6	1.1
4	7267.4	1.9	7235.7	1.9	7203.4	1.9
5	9084.3	2.9	9044.6	3.0	9004.3	3.0
6	10901.2	4.2	10853.6	4.3	10805.1	4.4
7	12718.0	5.8	12662.5	5.9	12606.0	5.9
8	14534.9	7.5	14471.4	7.6	14406.9	7.7
9	16351.7	9.5	16280.3	9.7	16207.7	9.8
10	18168.6	11.8	18089.3	11.9	18008.6	12.1
11	19985.5	14.3	19898.2	14.5	19809.4	14.6
12	21802.3	17.0	21707.1	17.2	21610.3	17.4
13	23619.2	19.9	23516.0	20.2	23411.1	20.4
14	25436.0	23.1	25325.0	23.4	25212.0	23.7
15	27252.9	26.5	27133.9	26.9	27012.8	27.2
16	29069.8	30.2	28942.8	30.6	28813.7	30.9
17	30886.6	34.1	30751.7	34.5	30614.6	34.9
18	32703.5	38.2	32560.7	38.7	32415.4	39.2
19	34520.3	42.6	34369.6	43.1	34216.3	43.6
20	36337.2	47.2	36178.5	47.8	36017.1	48.4
25	45421.4	73.7	45223.1	74.7	45021.4	75.6
30	54505.6	106.1	54267.7	107.5	54025.6	108.8
40	72674.1	188.6	72356.8	191.1	72034.0	193.5
50	90842.4	294.8	90445.8	298.6	90042.4	302.4
1° 00	109010.7	424.5	108534.8	430.0	108050.6	435.4
1 20	145346.7	754.6	144712.1	764.4	144066.5	774.0
1 30	163514.5	955.1	162800.5	967.5	162074.2	979.6
1 40	181682.0	1179.1	180888.7	1194.4	180081.7	1209.4
2 00	218016.4	1697.9	217064.4	1720.0	216095.9	1741.6
2 30	272515.9	2652.9	271325.7	2687.5	270114.9	2721.2
3 00	327012.2	3820.2	325583.8	3870.0	324130.7	3918.6
3 30	381505.0	5199.8	379838.2	5267.5	378142.5	5333.6
4 00	435993.2	6791.5	434088.0	6880.0	432149.7	6966.3

Co-ordinates, δm, δp, in Yards.

Long. value of Z.	Lat. 28° 0′		Lat. 28° 30′		Lat. 29°.	
	δm	δp	δm	δp	δm	δp
1′	1792.7	0.1	1784.3	0.1	1775.8	0.1
2	3585.3	0.5	3568.6	0.5	3551.7	0.5
3	5378.0	1.1	5352.9	1.1	5327.5	1.1
4	7170.6	2.0	7137.2	2.0	7103.3	2.0
5	8963.3	3.1	8921.5	3.1	8879.1	3.1
6	10755.9	4.4	10705.8	4.5	10655.0	4.5
7	12548.6	6.0	12490.2	6.1	12430.8	6.1
8	14341.2	7.8	14274.5	7.9	14206.6	8.0
9	16133.9	9.9	16058.8	10.0	15982.5	10.1
10	17926.5	12.2	17843.1	12.4	17758.3	12.5
11	19719.2	14.8	19627.4	15.0	19534.1	15.2
12	21511.8	17.6	21411.7	17.8	21309.9	18.0
13	23304.5	20.7	23196.0	20.9	23085.8	21.2
14	25097.1	24.0	24980.3	24.3	24861.6	24.5
15	26889.8	27.5	26764.6	27.9	26637.4	28.2
16	28682.4	31.3	28548.9	31.7	28413.2	32.1
17	30475.1	35.4	30333.2	35.8	30189.1	36.2
18	32267.7	39.7	32117.2	40.1	31964.9	40.6
19	34060.3	44.2	33901.8	44.7	33740.7	45.2
20	35853.0	49.0	35686.1	49.5	35516.6	50.1
25	44816.2	76.5	44607.6	78.4	44395.7	78.3
30	53779.4	110.1	53529.1	111.4	53274.8	112.7
40	71705.8	195.8	71372.1	198.1	71032.9	200.3
50	89632.0	306.0	89214.9	309.6	88790.9	313.0
1° 00	107558.2	440.7	107057.6	445.8	106548.8	450.8
1 20	143410.0	783.4	142742.5	792.5	142064.1	801.4
1 30	161335.6	991.5	160584.6	1003.0	159821.5	1014.3
1 40	179260.9	1224.2	178426.5	1238.3	177578.6	1252.2
2 00	215110.9	1762.6	214109.6	1783.2	213092.0	1803.1
2 30	268883.6	2754.1	267631.8	2786.2	266359.6	2817.4
3 00	322652.8	3965.9	321150.5	4012.1	319623.7	4057.1
3 30	376418.1	5398.1	374665.0	5460.9	372883.4	5522.1
4 00	430178.5	7050.6	428174.6	7132.7	426138.2	7212.6

Co-ordinates,　$\delta\,m$, $\delta\,p$, in Yards.

Long. value of Z.	Lat. 29° 30'		Lat. 30° 0'		Lat. 30° 30'	
	$\delta\,m$	$\delta\,p$	$\delta\,m$	$\delta\,p$	$\delta\,m$	$\delta\,p$
1'	1767.2	0.1	1758.5	0.1	1749.6	0.1
2	3534.4	0.5	3516.9	0.5	3499.2	0.5
3	5301.6	1.1	5275.4	1.2	5248.8	1.2
4	7068.9	2.0	7033.9	2.0	6998.3	2.1
5	8836.1	3.2	8792.3	3.2	8747.9	3.2
6	10603.3	4.6	10550.8	4.6	10497.5	4.6
7	12370.5	6.2	12309.3	6.3	12247.1	6.3
8	14137.7	8.1	14067.8	8.2	13996.7	8.2
9	15904.9	10.3	15826.2	10.4	15746.3	10.5
10	17672.7	12.7	17584.7	12.8	17495.9	12.9
11	19439.4	15.3	19343.1	15.5	19245.4	15.6
12	21206.6	18.2	21101.6	18.4	20995.0	18.6
13	22973.8	21.4	22860.1	21.6	22744.6	21.8
14	24741.0	24.8	24618.5	25.1	23494.2	25.3
15	26508.2	28.5	26377.0	28.8	25243.8	29.1
16	28275.4	32.4	28135.5	32.7	26993.3	33.1
17	30042.6	36.6	29893.9	37.0	28742.9	37.4
18	31809.9	41.0	31652.4	41.4	30492.5	41.8
19	33577.1	45.7	33410.9	46.2	32242.1	46.6
20	35344.3	51.6	35169.3	51.2	33991.7	51.7
25	44180.3	79.1	43961.6	79.9	43239.6	80.7
30	53016.4	113.9	52753.9	115.1	52487.4	116.3
40	70688.4	202.5	70338.4	204.6	69983.1	206.6
50	88360.2	316.4	87922.8	319.7	87478.7	322.9
1° 00	106032.0	455.6	105507.1	460.4	104974.1	464.9
1 20	141375.0	810.0	140675.1	818.4	139964.4	826.6
1 30	159046.2	1025.2	158258.7	1035.8	157459.2	1046.1
1 40	176717.1	1265.7	175842.2	1278.8	174953.8	1291.5
2 00	212058.1	1822.6	211008.1	1841.5	209942.1	1859.8
2 30	265067.1	2847.8	263754.5	2877.3	262421.8	2905.9
3 00	318072.5	4100.8	316497.1	4143.3	314897.6	4184.5
3 30	371073.4	5581.7	369235.2	5639.5	367368.9	5695.6
4 00	424069.3	7290.3	421968.0	7365.9	419834.7	7439.1

Co-ordinates, $\delta\,m$, $\delta\,p$, in Yards.

Long. value of Z.	Lat. 31° 0'.		Lat. 31° 30'		Lat. 32° 0'	
	$\delta\,m$	$\delta\,p$	$\delta\,m$	$\delta\,p$	$\delta\,m$	$\delta\,p$
1'	1740.6	0.1	1731.4	0.1	1722.1	0.1
2	3481.1	0.5	3462.8	0.5	3444.3	0.5
3	5221.7	1.2	5194.3	1.2	5166.4	1.2
4	6962.3	2.1	6925.7	2.1	6888.6	2.1
5	8702.9	3.3	8657.1	3.3	8610.7	3.3
6	10443.4	4.7	10388.5	4.7	10332.8	4.8
7	12184.0	6.4	12119.9	6.4	12055.0	6.5
8	13924.6	8.3	13851.4	8.4	13777.1	8.5
9	15665.1	10.6	15582.8	10.7	15499.3	10.8
10	17405.7	13.0	17314.2	13.2	17221.4	13.3
11	19146.3	15.8	19045.6	15.9	18943.6	16.1
12	20886.8	18.8	20777.1	18.9	20665.7	19.1
13	22627.4	22.0	22508.5	22.2	22387.8	22.4
14	24368.0	25.6	24239.9	25.8	24110.0	26.0
15	26108.5	29.3	25971.3	29.6	25832.1	29.9
16	27849.1	33.4	27702.7	33.7	27554.3	34.0
17	29589.7	37.7	29434.2	38.0	29276.4	38.4
18	31330.3	42.2	31165.6	42.6	30998.5	43.0
19	33070.8	47.1	32897.0	47.5	32720.7	47.9
20	34811.4	52.2	34628.4	52.6	34442.8	53.1
25	43514.2	81.5	43286.0	82.3	43053.5	83.0
30	52217.0	117.3	51942.5	118.4	51664.1	119.5
40	69622.5	208.6	69256.6	210.5	68885.4	212.4
50	87027.9	326.0	86570.5	328.9	86106.5	331.8
1° 00	104433.2	469.4	103884.3	473.7	103327.4	477.8
1 20	139243.1	834.5	138511.2	842.1	137768.8	849.5
1 30	156647.8	1056.1	155824.4	1065.8	154989.1	1075.1
1 40	174052.2	1303.8	173137.2	1315.8	172209.1	1327.3
2 00	208860.0	1877.5	207762.0	1894.7	206648.2	1911.3
2 30	261069.1	2933.7	259696.5	2960.5	258304.1	2986.5
3 00	313274.2	4224.5	311626.9	4263.1	309955.8	4300.5
3 30	365474.6	5750.0	363552.4	5802.6	361602.5	5853.5
4 00	417669.4	7510.2	415472.3	7578.9	413243.4	7645.3

Co-ordinates, δ m, δ p, in Yards.

Long. value of Z.	Lat 32° 30'		Lat. 33° 0'		Lat. 33° 30'	
	δ m	δ p	δ m	δ p	δ m	p
1′	1712.7	0.1	1703.2	0.1	1693.5	0.1
2	3425.5	0.5	3406.4	0.5	3387.0	0.5
3	5138.2	1.2	5109.6	1.2	5080.6	1.2
4	6850.9	2.1	6812.8	2.2	6774.1	2.2
5	8563.7	3.3	8515.9	3.4	8467.6	3.4
6	10276.4	4.8	10219.1	4.9	10161.1	4.9
7	11989.1	6.6	11922.3	6.6	18154.6	6.7
8	13701.8	8.6	13625.5	8.6	13548.1	8.7
9	15414.6	10.8	15328.7	10.9	15241.7	11.0
10	17127.3	13.4	17031.9	13.5	16935.2	13.6
11	18840.0	16.2	18735.1	16.3	18628.7	16.4
12	20552.8	19.3	20438.3	19.4	20322.2	19.6
13	22265.5	22.6	22141.4	22.8	22015.7	23.0
14	23978.2	26.2	23844.6	26.4	23709.2	26.6
15	25690.9	30.1	25547.8	30.4	25402.7	30.6
16	27403.7	34.3	27251.0	34.5	27096.3	34.9
17	29116.4	38.7	28954.2	39.0	28789.8	39.3
18	30829.1	43.4	30657.4	43.7	30483.3	44.0
19	32541.9	48.5	32360.6	49.0	32176.8	49.1
20	34254.6	53.5	34063.8	54.0	33870.3	54.4
25	42818.2	83.6	42579.6	84.3	42337.9	85.0
30	51381.8	120.5	51095.5	121.4	50805.4	122.3
40	68508.9	214.1	68127.2	215.9	67740.4	217.5
50	85635.9	334.6	85158.8	337.3	84675.2	339.9
1° 00	102762.7	481.8	102190.2	485.7	101609.9	489.4
1 20	137015.8	856.6	136252.4	863.5	135478.6	870.1
1 30	154142.0	1084.1	153283.1	1092.8	152412.6	1101.2
1 40	171267.0	1338.4	170313.6	1349.2	169346.3	1359.5
2 00	205518.7	1927.4	204373.5	1942.8	203212.7	1957.7
2 30	256892.0	3011.5	255460.4	3035.6	254009.2	3058.8
3 00	308261.1	4336.6	306542.9	4371.3	304801.4	4404.7
3 30	359625.1	5902.6	357620.3	5949.9	355588.2	5995.3
4 00	410983.2	7709.5	408691.6	7771.2	406368.8	7830.6

Co-ordinates, δm, δp, in Yards.

Long. value of Z.		Lat. 34° 0′		Lat. 34° 30′		Lat. 35° 0′	
		δm	δp	δm	δp	δm	δp
	1′	1683.7	0.1	1673.8	0.1	1663.7	0.1
	2	3367.4	0.5	3347.6	0.6	3327.5	0.6
	3	5051.1	1.2	5021.4	1.2	4991.2	1.2
	4	6734.9	2.2	6695.1	2.2	6654.9	2.2
	5	8418.6	3.4	8368.9	3.4	8318.7	3.5
	6	10102.3	4.9	10042.7	5.0	9982.4	5.0
	7	11786.0	6.7	11716.5	6.8	11646.1	6.8
	8	13469.7	8.8	13390.3	8.8	13309.8	8.9
	9	15153.4	11.1	15064.1	11.2	14973.6	11.2
	10	16837.2	13.7	16737.9	13.8	16637.3	13.9
	11	18520.9	16.6	18411.6	16.7	18301.0	16.8
	12	20204.6	19.7	20085.4	19.9	19964.8	20.0
	13	21888.3	23.1	21759.2	23.3	21628.5	23.5
	14	23572.0	26.8	23433.0	27.0	23292.2	27.2
	15	25255.7	30.8	25106.8	31.0	24955.9	31.2
	16	26939.4	35.1	26780.6	35.3	26619.7	35.5
	17	28623.1	39.6	28454.4	39.8	28283.4	40.1
	18	30306.9	44.4	30128.1	44.7	29947.1	45.0
	19	31990.6	49.4	31801.9	49.8	31610.9	50.1
	20	33674.3	54.8	33475.7	55.2	33274.6	55.5
	25	42092.8	85.6	41844.6	86.2	41593.2	86.7
	30	50511.4	123.2	50213.5	125.1	49911.8	124.9
	40	67348.3	219.1	66951.1	220.6	65548.9	222.1
	50	84185.1	342.3	83688.7	344.7	83185.8	347.0
1°	00	101021.8	493.0	100426.0	496.4	99822.6	499.7
1	20	134694.5	876.4	133900.1	882.5	133095.5	883.3
1	30	151530.5	1109.2	150636.8	1116.9	149731.5	1124.2
1	40	168366.1	1369.4	167373.1	1378.9	166367.3	1387.9
2	00	202036.4	1971.9	200844.7	1985.6	199637.8	1998.6
2	30	252538.7	3081.1	251049.0	3102.5	249540.1	3122.8
3	00	303036.6	4436.8	301248.7	4467.5	299437.8	4496.9
3	30	353529.0	6039.0	351442.8	6080.8	349329.8	6120.8
4	00	404015.1	7887.7	401630.5	7942.3	399215.4	7994.5

Co-ordinates, δm, δp, in Yards.

Long. value of Z.	Lat. 35° 30′		Lat. 36° 0′		Lat. 36° 30′	
	δm	δp	δm	δp	δm	δp
1′	1653.5	0.1	1643.2	0.1	1632.8	0.1
2	3307.1	0.6	3286.5	0.6	3265.6	0.6
3	4960.6	1.3	4929.7	1.3	4898.4	1.3
4	6614.2	2.2	6572.9	2.2	6531.2	2.3
5	8267.7	3.5	8216.2	3.5	8164.0	3.5
6	9921.3	5.0	9859.4	5.1	9796.8	5.1
7	11574.8	6.8	11502.7	6.9	11429.6	6.9
8	13228.4	8.9	13145.9	9.0	13062.4	9.0
9	14881.9	11.3	14789.1	11.4	14695.2	11.4
10	16535.5	14.0	16432.4	14.0	16328.0	14.1
11	18189.0	16.9	18075.6	17.0	17960.8	17.0
12	19842.5	20.1	19718.8	20.2	19593.6	20.3
13	21496.1	23.6	21362.0	23.7	21226.4	23.9
14	23149.6	27.4	23005.3	27.5	22859.2	27.7
15	24803.2	31.4	24648.5	31.6	24492.0	31.8
16	26456.7	35.7	26291.8	36.0	26124.8	36.2
17	28110.3	40.4	27935.0	40.6	27757.6	40.8
18	29763.8	45.2	29578.2	45.5	29390.4	45.8
19	31417.3	50.4	31221.2	50.7	31023.2	51.0
20	33070.9	55.9	32864.7	56.2	32656.0	56.5
25	41338.6	87.2	41080.8	87.8	40819.9	88.3
30	49606.2	125.7	49296.9	126.4	48983.9	127.1
40	66141.5	223.4	65729.1	224.8	65311.7	226.0
50	82676.6	349.1	82161.1	351.2	81639.3	353.1
1° 00	99211.5	502.7	98592.9	507.7	97966.7	508.5
1 20	132280.7	893.8	131455.9	899.1	130621.0	904.1
1 30	148814.9	1131.2	147886.9	1137.9	146947.7	1144.2
1 40	165348.8	1396.6	164317.7	1404.8	163274.0	1412.6
2 00	198415.4	2011.1	197178.0	2022.9	195925.6	2034.1
2 30	248012.1	3142.3	246465.3	3160.8	244899.6	3178.3
3 00	297604.0	4524.9	295747.6	4551.6	293868.6	4576.8
3 30	347190.2	6158.9	345024.1	6195.2	342831.7	6229.5
4 00	396769.7	8044.3	394293.8	8091.6	391787.8	8136.5

Co-ordinates, $\delta\, m$, $\delta\, p$, in Yards.

Long. value of Z.	Lat. 37° 0′		Lat. 37° 30′		Lat. 38° 0′	
	$\delta\, m$	$\delta\, p$	$\delta\, m$	$\delta\, p$	$\delta\, m$	$\delta\, p$
1′	1622.2	0.1	1611.6	0.1	1600.7	0.1
2	3244.5	0.6	3223.1	0.6	3201.5	0.6
3	4866.7	1.3	4834.7	1.3	4802.2	1.3
4	6489.0	2.3	6446.2	2.3	6403.0	2.3
5	8111.2	3.5	8057.8	3.6	8003.7	3.6
6	9733.4	5.1	9669.3	5.1	9604.5	5.2
7	11355.7	7.0	11280.9	7.0	11205.2	7.0
8	12977.9	9.1	12892.4	9.1	12806.0	9.2
9	14600.2	11.5	14504.0	11.6	14406.7	11.6
10	16222.4	14.2	16115.6	14.3	16007.5	14.3
11	17844.6	17.2	17727.1	17.3	17608.2	17.3
12	19466.9	20.4	19338.6	20.5	19209.0	20.6
13	21089.1	24.0	20950.2	24.1	20809.7	24.2
14	22711.4	27.8	22561.8	28.0	22410.5	28.1
15	24333.6	31.9	24173.3	32.1	24011.2	32.3
16	25955.8	36.4	25784.9	36.5	25612.0	36.7
17	27578.1	41.0	27396.4	41.2	27212.7	41.4
18	29200.3	46.0	29008.0	46.2	28813.4	46.4
19	30822.6	51.3	30619.5	51.5	30414.2	51.7
20	32444.8	56.8	32231.1	57.1	32015.0	57.3
25	40555.9	88.7	40288.8	89.1	40018.6	89.6
30	48667.1	127.8	48346.5	128.4	48022.3	129.0
40	64889.2	227.2	64461.9	228.3	64029.6	229.3
50	81111.3	355.0	80577.1	356.7	80036.7	358.3
1° 00	97333.1	511.2	96692.0	513.7	96043.6	516.0
1 20	129776.1	908.8	128921.3	913.2	128056.7	917.4
1 30	145997.2	1150.2	145035.5	1155.8	144062.6	1161.0
1 40	162217.9	1419.9	161149.4	1426.9	160068.5	1433.4
2 00	194658.2	2044.7	193375.9	2054.7	192078.9	2064.1
2 30	243315.2	3194.9	241712.2	3210.5	240090.8	3225.1
3 00	291967.1	4600.7	290043.4	4623.1	288097.5	4644.1
3 30	340613.1	6262.0	338368.4	6292.6	336098.0	6321.2
4 00	389251.9	8178.9	386686.3	8218.8	384091.9	8256.3

Co-ordinates, δm, δp, in Yards.

Long. value of Z.	Lat. 38° 30′		Lat. 39° 0′		Lat. 39° 30′	
	δm	δp	δm	δp	δm	δp
1′	1589.8	0.1	1578.8	0.1	1567.6	0.1
2	3179.6	0.6	3157.5	0.6	3135.2	0.6
3	4769.5	1.3	4736.3	1.3	4702.8	1.3
4	6359.3	2.3	6315.1	2.3	6270.4	2.3
5	7949.1	3.6	7893.8	3.6	7838.0	3.6
6	9538.9	5.2	9472.6	5.2	9405.6	5.2
7	11128.7	7.1	11051.4	7.1	10973.2	7.1
8	12718.6	9.2	12630.1	9.2	12540.8	9.3
9	14308.4	11.7	14208.9	11.7	14108.4	11.7
10	15898.2	14.4	15787.7	14.5	15676.0	14.5
11	17488.0	17.4	17366.4	17.5	17243.6	17.5
12	19077.8	20.7	18945.2	20.8	18811.1	20.9
13	20667.6	24.3	20524.0	24.4	20378.7	24.5
14	22257.4	28.2	22102.7	28.3	21946.3	28.4
15	23847.3	32.4	23681.5	32.5	23513.9	32.6
16	25437.1	36.8	25260.3	37.0	25081.5	37.1
17	27026.9	41.6	26839.0	41.8	26649.1	41.9
18	28616.7	46.6	28417.8	46.8	28216.7	47.0
19	30206.5	52.0	29996.6	52.2	29784.3	52.3
20	31796.3	57.6	31575.3	57.8	31351.9	58.0
25	39745.4	90.0	39469.1	90.3	39189.8	90.6
30	47694.4	129.5	47362.9	130.1	47027.7	130.5
40	63592.4	230.3	63150.3	231.2	62703.4	232.0
50	79490.2	359.9	78937.6	361.3	78379.0	362.6
1° 00	95387.8	518.2	94724.7	520.2	94054.4	522.1
1 20	127182.3	921.2	126298.1	924.8	125404.3	928.2
1 30	143079.0	1165.9	142084.4	1170.5	141078.8	1174.7
1 40	158975.5	1439.4	157870.3	1445.1	156753.0	1450.2
2 00	190767.1	2072.8	189440.8	2080.9	188100.1	2088.3
2 30	238451.0	3238.7	236793.0	3251.4	235116.9	3263.0
3 00	286129.6	4663.8	284139.8	4682.0	282128.3	4698.8
3 30	333801.8	6347.9	331480.2	6372.7	329133.2	6395.6
4 00	381466.7	8291.2	378813.1	8323.6	376130.5	8353.4

Co-ordinates, δm, δp, in Yards.

Long. value of Z.	Lat. 40° 0′		Lat. 40° 30′		Lat. 41° 0′	
	δm	δp	δm	δp	δm	δp
1′	1556.3	0.1	1544.9	0.1	1533.4	0.1
2	3112.6	0.6	3089.8	0.6	3066.7	0.6
3	4668.9	1.3	4634.7	1.3	4600.1	1.3
4	6225.2	2.3	6179.6	2.3	6133.5	2.3
5	7781.5	3.6	7724.5	3.6	7666.8	3.7
6	9337.8	5.2	9269.4	5.3	9200.2	5.3
7	10894.1	7.1	10814.3	7.2	10733.6	7.2
8	12450.4	9.3	12359.2	9.3	12266.9	9.4
9	14006.7	11.8	13904.0	11.8	13800.3	11.9
10	15563.0	14.6	15448.9	14.6	15333.7	14.6
11	17119.3	17.6	16993.8	17.7	16867.0	17.7
12	18675.6	21.0	18538.7	21.0	18400.4	21.4
13	20231.9	24.6	20083.6	24.7	19933.7	24.7
14	21788.2	28.5	21628.5	28.6	21467.1	28.7
15	23344.5	32.7	23173.4	32.8	23000.4	32.9
16	24900.8	37.2	24718.3	37.4	24533.8	37.5
17	26457.1	42.0	26263.2	42.2	26067.2	42.3
18	28013.4	47.1	27808.1	47.3	27600.5	47.4
19	29569.8	52.5	29352.9	52.7	29133.9	52.8
20	31126.1	58.2	30897.8	58.4	30667.3	58.5
25	38907.5	90.9	38622.2	91.2	38334.0	91.4
30	46689.0	130.9	46346.7	131.3	46000.8	131.7
40	62251.8	232.8	61795.3	233.5	61334.2	234.1
50	77814.4	363.7	77243.9	364.8	76667.4	365.8
1° 00	93376.9	523.8	92692.2	525.3	92000.4	526.7
1 20	124500.9	931.2	123588.0	933.9	122665.7	936.4
1 30	140062.5	1178.5	139035.5	1182.0	137997.8	1185.1
1 40	155623.7	1455.0	154482.6	1459.3	153329.6	1463.1
2 00	186744.9	2095.2	185375.4	2101.4	183991.8	2106.9
2 30	233422.9	3273.7	231710.9	3283.4	229998.3	3292.0
3 00	280095.3	4714.1	278040.9	4728.1	275965.1	4740.6
3 30	326761.1	6416.5	324364.1	6435.4	321942.2	6452.4
4 00	373419.3	8380.7	370679.4	8405.5	367911.3	8427.7

Co-ordinates, $\delta\,m,\ \delta\,p,$ *in Yards.*

Long. value of Z.	Lat. 41° 30′		Lat. 42° 0′		Lat. 42° 30′	
	$\delta\,m$	$\delta\,p$	$\delta\,m$	$\delta\,p$	$\delta\,m$	$\delta\,p$
1	1521.7	0.1	1510.0	0.1	1498.1	0.1
2	3043.4	0.6	3019.9	0.6	2996.2	0.6
3	4565.2	1.3	4529.9	1.3	4494.2	1.3
4	6086.9	2.3	6039.8	2.4	5992.3	2.4
5	7608.6	3.7	7549.8	3.7	7490.4	3.7
6	9130.3	5.3	9059.7	5.3	8988.6	5.3
7	10652.0	7.2	10569.7	7.2	10486.5	7.2
8	12173.8	9.4	12079.6	9.4	11984.6	9.4
9	13695.5	11.9	13589.6	11.9	13482.7	11.9
10	15217.2	14.7	15099.6	14.7	14980.8	14.7
11	16738.9	17.7	16609.5	17.8	16478.8	17.8
12	18260.6	21.1	18119.5	21.2	17976.9	21.2
13	19782.3	24.8	19629.4	24.8	19475.0	24.9
14	21304.0	28.7	21139.4	28.8	20973.1	28.8
15	22825.8	33.0	22649.3	33.1	22471.1	33.1
16	24347.5	37.5	24159.3	37.6	23969.2	37.6
17	25869.2	42.4	25669.2	42.5	25467.3	42.5
18	27390.9	47.5	27179.2	47.6	26965.4	47.7
19	28912.6	52.9	28689.1	53.0	28463.4	53.1
20	30434.3	58.6	30199.1	58.8	29961.5	58.9
25	38042.9	91.6	37748.8	91.8	37451.3	92.0
30	45651.4	132.0	45298.5	132.3	44942.2	132.5
40	60868.3	234.6	60397.8	235.1	59922.7	235.5
50	76085.1	366.6	75496.9	367.4	74903.0	368.0
1° 00	91301.6	527.9	90595.9	529.0	89883.2	529.9
1 20	121733.9	938.6	120792.9	940.5	119842.6	942.1
1 30	136949.6	1187.9	135890.9	1190.3	134821.8	1192.3
1 40	152164.9	1466.5	150988.5	1469.5	149800.6	1472.0
2 00	182594.1	2111.8	181182.5	2116.1	179756.9	2119.7
2 30	228234.1	3299.7	226469.4	3306.4	224687.4	3312.0
3 00	273868.3	4751.6	271750.5	4761.2	269612.0	4769.3
3 30	319495.6	6467.5	317024.7	6480.5	314529.5	6491.6
4 00	365115.0	8447.3	362290.8	8464.4	359438.9	8478.8

Co-ordinates, $\delta\, m$, $\delta\, p$, in Yards.

Long. value of Z.	Lat. 43° 0′		Lat. 43° 30′		Lat. 44° 0′	
	$\delta\, m$	$\delta\, p$	$\delta\, m$	$\delta\, p$	$\delta\, m$	$\delta\, p$
1′	1486.1	0.1	1474.0	0.1	1461.8	0.1
2	2972.2	0.6	2948.0	0.6	2923.5	0.6
3	4458.3	1.3	4421.9	1.3	4385.3	1.3
4	5944.3	2.4	5895.9	2.4	5847.0	2.4
5	7430.4	3.7	7369.9	3.7	7308.8	3.7
6	8916.5	5.3	8843.9	5.3	8770.5	5.3
7	10402.6	7.2	10317.8	7.2	10232.3	7.2
8	11888.7	9.4	11791.8	9.5	11694.1	9.5
9	13374.8	11.9	13265.8	12.0	13155.8	12.0
10	14860.8	14.7	14739.8	14.8	14617.6	14.8
11	16346.9	17.8	16213.7	17.8	16079.3	17.9
12	17833.0	21.2	17687.7	21.2	17541.1	21.3
13	19319.1	24.9	19161.7	24.9	19002.8	25.0
14	20805.2	28.9	20635.7	28.9	20464.6	28.9
15	22291.2	33.2	22109.6	33.2	21926.3	33.2
16	23777.3	37.7	23583.6	37.8	23388.1	37.8
17	25263.4	42.6	25057.6	42.6	24849.9	42.7
18	26749.5	47.8	26531.6	47.8	26311.6	47.9
19	28235.6	53.2	28005.5	53.3	27773.4	53.3
20	29721.7	59.0	29479.5	59.0	29235.1	59.1
25	37102.0	92.1	36849.3	92.2	36543.8	92.3
30	44582.4	132.7	44219.2	132.8	43852.6	132.9
40	59443.0	235.8	58958.7	236.1	58469.9	236.3
50	74303.4	368.5	73698.0	368.9	73087.0	369.2
1° 00	89163.6	530.7	88437.1	531.2	87703.9	531.7
1 20	118883.1	943.4	117914.5	944.4	116936.9	945.2
1 30	133742.4	1194.0	132652.7	1195.3	131552.9	1196.3
1 40	148601.2	1474.1	147390.5	1475.7	146168.4	1476.9
2 00	178317.6	2122.7	176864.7	2125.0	175398.2	2126.7
2 30	222888.2	3316.7	221071.9	3320.3	219238.7	3323.0
3 00	267452.8	4776.0	265273.1	4781.3	263073.1	4785.1
3 30	312010.3	6500.7	309467.1	6507.9	306900.3	6513.0
4 00	356559.5	8490.7	353652.8	8500.1	350719.0	8506.8

15

Co-ordinates, $\delta\,m$, $\delta\,p$, *in Yards.*

Long. value of Z.	Lat. 44° 30′		Lat. 45° 0′		Lat. 45° 30′	
	$\delta\,m$	$\delta\,p$	$\delta\,m$	$\delta\,p$	$\delta\,m$	$\delta\,p$
1′	1449.4	0.1	1437.0	0.1	1424.4	0.1
2	2898.9	0.6	2874.0	0.6	2848.9	0.6
3	4348.3	1.3	4311.0	1.3	4273.3	1.3
4	5797.7	2.4	5747.9	2.4	5697.7	2.4
5	7247.1	3.7	7184.9	3.7	7122.2	3.7
6	8696.6	5.3	8621.9	5.3	8546.6	5.3
7	10146.0	7.2	10058.9	7.2	9971.0	7.2
8	11595.4	9.5	11495.9	9.5	11395.4	9.5
9	13044.8	12.0	12932.9	12.0	12819.9	12.0
10	14494.3	14.8	14369.8	14.8	14244.3	14.8
11	15943.7	17.9	15806.8	17.9	15668.7	17.9
12	17393.1	21.3	17243.8	21.3	17093.1	21.3
13	18842.5	25.0	18680.8	25.0	18517.6	25.0
14	20292.0	29.0	20117.7	29.0	19942.0	29.0
15	21741.4	33.2	21554.7	33.3	21366.4	33.2
16	23190.8	37.8	22991.7	37.8	22790.8	37.8
17	24640.2	42.7	24428.7	42.7	24215.3	42.7
18	26089.7	47.9	25865.7	47.9	25639.7	47.9
19	27539.1	53.3	27302.7	53.4	27064.1	53.3
20	28988.5	59.1	28739.6	59.1	28488.6	59.1
25	36235.6	92.3	35924.5	92.4	35610.6	92.3
30	43482.6	133.0	43109.3	133.0	42732.7	133.0
40	57976.6	236.4	57478.9	236.5	56976.8	236.4
50	72470.4	369.4	71848.3	369.5	71220.6	369.4
1° 00	86964.0	531.9	86217.4	532.0	85464.2	532.0
1 20	115950.3	945.7	114954.9	945.8	113950.6	945.7
1 30	130442.9	1196.8	129323.0	1197.1	128193.2	1196.9
1 40	144935.2	1477.6	143690.8	1477.9	142435.5	1477.7
2 00	173918.3	2127.7	172425.0	2128.1	170918.5	2127.8
2 30	217388.7	3324.6	215522.0	3325.2	213638.8	3324.8
3 00	260853.0	4787.4	258612.9	4788.3	256352.9	4787.7
3 30	304310.0	6516.2	301696.3	6517.3	299059.6	6516.5
4 00	347758.4	8510.9	344771.2	8512.5	341757.6	8511.4

Co-ordinates, δm, δp, in Yards.

Long. value of Z.	Lat. 46° 0′		Lat. 46° 30′		Lat. 47° 0′	
	δm	δp	δm	δp	δm	δp
1′	1411.8	0.1	1399.0	0.1	1386.1	0.1
2	2823.5	0.6	2798.0	0.6	2772.2	0.6
3	4235.3	1.3	4197.0	1.3	4158.4	1.3
4	5647.1	2.4	5596.0	2.4	5544.5	2.4
5	7058.8	3.7	6995.0	3.7	6930.6	3.7
6	8470.6	5.3	8394.0	5.3	8316.7	5.3
7	9882.4	7.2	9793.0	7.2	9702.8	7.2
8	11294.1	9.5	11192.0	9.4	11089.0	9.4
9	12705.9	12.0	12591.0	12.0	12475.1	11.9
10	14117.7	14.8	13990.0	14.8	13861.2	14.7
11	15529.4	17.9	15389.0	17.9	15247.3	17.8
12	16941.2	21.3	16788.0	21.3	16633.4	21.2
13	18353.0	25.0	18186.9	24.9	18019.5	24.9
14	19764.7	28.9	19585.9	28.9	19405.7	28.9
15	21176.5	33.2	20984.9	33.2	20791.8	33.2
16	22588.3	37.8	22383.9	37.8	22177.9	37.7
17	24000.0	42.7	23782.9	42.7	23564.0	42.6
18	25411.8	47.9	25181.9	47.8	24950.1	47.8
19	26823.6	53.3	26580.9	53.3	26336.2	53.2
20	28235.3	59.1	27979.9	59.0	27722.4	59.0
25	35294.1	92.3	24974.8	92.2	34652.9	92.2
30	42352.9	132.9	41969.8	132.8	41583.4	132.7
40	56470.3	236.3	55959.5	236.2	55444.3	235.9
50	70587.5	369.3	69949.0	369.0	69305.1	368.6
1° 00	84704.5	531.7	83938.2	531.3	83165.6	530.8
1 20	112937.6	945.3	111915.9	944.6	110885.7	943.6
1 30	127053.6	1196.4	125904.2	1195.5	124747.2	1194.3
1 40	141169.2	1477.0	139892.1	1476.0	138604.3	1474.4
2 00	169399.0	2126.9	167866.4	2125.4	166321.0	2123.2
2 30	211739.3	3323.3	209823.5	3320.9	207891.7	3317.5
3 00	254073.4	4785.6	251774.4	4782.1	249456.0	4777.2
3 30	296400.0	6513.8	293717.6	6509.0	291012.8	6502.2
4 00	338717.8	8507.8	335652.1	8501.5	332560.6	8492.7

Co-ordinates, $\delta\,m$, $\delta\,p$, in Yards,

Long. value of Z.	Lat. 47° 30′		Lat. 48° 0′		Lat. 48° 30′	
	$\delta\,m$	$\delta\,p$	$\delta\,m$	$\delta\,p$	$\delta\,m$	$\delta\,p$
1′	1373.1	0.1	1360.0	0.1	1346.9	0.1
2	2746.3	0.6	2720.1	0.6	2693.7	0.6
3	4119.4	1.3	4080.1	1.3	4040.6	1.3
4	5492.5	2.4	5440.2	2.4	5387.4	2.4
5	6865.7	3.7	6800.2	3.7	6734.3	3.7
6	8238.8	5.3	8160.3	5.3	8081.1	5.3
7	9612.0	7.2	9520.3	7.2	9428.0	7.2
8	10985.1	9.4	10880.4	9.4	10774.8	9.4
9	12358.2	11.9	12240.4	11.9	12121.7	11.9
10	13731.4	14.7	13600.5	14.7	13468.5	14.7
11	15104.5	17.8	14960.5	17.8	14815.4	17.8
12	16477.6	21.2	16320.5	21.2	16162.2	21.1
13	17850.8	24.9	17680.6	24.8	17509.1	24.8
14	19223.9	28.9	19040.6	28.8	18855.9	28.8
15	20597.0	33.1	20400.7	33.1	20202.8	33.0
16	21970.1	37.7	21760.7	37.6	21549.6	37.6
17	23343.3	42.6	23120.7	42.5	22896.5	42.4
18	24716.4	47.7	24480.8	47.6	24243.3	47.5
19	26089.5	53.1	25840.8	53.1	25590.2	53.0
20	27462.7	58.9	27200.9	58.8	26937.0	58.7
25	34328.3	92.0	34001.0	91.9	33671.2	91.7
30	41193.9	132.5	40801.2	132.3	40405.4	132.0
40	54925.0	235.6	54401.4	235.2	53873.6	234.7
50	68655.8	368.1	68001.4	367.5	67341.7	366.8
1° 00	82386.5	530.1	81601.1	529.2	80809.5	528.2
1 20	109846.9	942.4	108799.7	940.8	107744.2	939.0
1 30	123576.6	1192.7	122398.5	1190.7	121211.0	1188.4
1 40	137305.6	1472.4	135996.8	1470.0	134677.3	1467.1
2 00	164762.8	2120.3	163191.9	2116.8	161608.6	2112.7
2 30	205943.9	3313.0	203980.3	3307.5	202001.0	3301.1
3 00	247118.6	4770.7	244762.2	4762.9	242387.0	4753.5
3 30	288285.6	6493.5	285536.3	6482.8	282765.2	6470.1
4 00	329443.7	8481.3	326301.5	8467.3	323134.3	8450.7

Co-ordinates, $\delta.m$, δp, in Yards.

Long. value of Z.	Lat. 49° 0′		Lat. 49° 30′		Lat. 50° 0′	
	δm	δp	δm	δp	δm	δp
1′	1333.6	0.1	1320.2	0.1	1306.7	0.1
2	2667.1	0.6	2640.3	0.6	2213.3	0.6
3	4000.7	1.3	2960.5	1.3	3920.0	1.3
4	5334.2	2.3	5280.6	2.3	5226.6	2.3
5	6667.8	3.7	6600.8	3.7	6533.3	3.6
6	8001.3	5.3	7920.9	5.3	7839.9	5.2
7	9334.9	7.2	9241.1	7.2	9146.6	7.1
8	10668.4	9.4	10561.2	9.3	10453.2	9.3
9	12002.0	11.9	11881.4	11.8	11759.9	11.8
10	13335.5	14.6	13201.6	14.6	13066.5	14.6
11	14669.1	17.7	14521.7	17.7	14373.2	17.6
12	16002.7	21.1	15841.8	21.0	15679.8	21.0
13	17336.2	24.7	17162.0	24.7	16986.5	24.6
14	18669.7	28.7	18482.1	28.6	18293.1	28.5
15	20003.3	32.9	19802.3	32.8	19599.8	32.8
16	21336.8	37.5	21122.4	37.4	20906.4	37.3
17	22670.4	42.3	22442.6	42.2	22213.1	42.1
18	24004.0	47.4	23762.8	47.3	23519.7	47.2
19	25337.5	52.8	25082.9	52.7	24826.4	52.6
20	26671.1	58.6	26403.1	58.4	26133.0	58.2
25	33338.8	91.5	33003.8	91.2	32666.2	91.0
30	40006.5	131.7	39604.5	131.4	39199.4	131.0
40	53341.7	234.2	52805.7	233.6	52265.7	232.9
50	66676.8	365.9	66006.8	365.0	65331.7	364.0
1° 00	80011.6	527.0	79207.6	525.6	78397.5	524.1
1 20	106680.4	936.8	105608.3	934.4	104528.2	931.7
1 30	120014.2	1185.7	118808.2	1182.6	117593.0	1179.2
1 40	133347.5	1463.8	132007.5	1460.0	130657.4	1455.8
2 00	160012.8	2107.9	158404.8	2102.5	156784.7	2096.4
2 30	200006.3	3293.6	197996.2	3285.1	195970.8	3275.6
3 00	239993.2	4742.8	237581.0	4730.5	235150.6	4716.9
3 30	279972.3	6455.4	277158.0	6438.8	274322.4	6420.2
4 00	319942.3	4831.6	316725.8	8409.9	313485.0	8385.6

Arcs of the Parallel—Values of D *p, in Yards.*

′	″	L. 20° 30′	L. 21° 0′	L. 21° 30′	L. 22° 0′	L. 22° 30′	L. 23° 0′
	7	221.8	221.1	220.3	219.6	218.8	218.0
	8	253.5	252.6	251.8	250.9	250.0	249.1
	9	285.2	284.2	283.3	282.3	281.3	280.3
	10	316.9	315.8	314.7	313.7	312.5	311.4
	20	633.7	631.6	629.5	627.3	625.1	622.8
	30	950.6	947.4	944.2	941.0	937.6	934.2
	40	1267.4	1263.2	1259.0	1254.6	1250.2	1245.7
	50	1584.3	1579.1	1573.7	1568.3	1562.7	1557.1
	60	1901.1	1894.9	1888.5	1881.9	1875.3	1868.5
7	–	13307.8	13264.1	13219.4	13173.7	13127.0	13079.4
8	–	15208.9	15159.0	15107.9	15055.7	15002.3	14947.9
9	–	17110.0	17053.8	16996.4	16937.6	16877.6	16816.3
10	–	19011.1	18948.7	18884.9	18819.6	18752.9	18684.8
20	–	38022.1	37897.4	37769.7	37639.2	37505.8	37369.6
30	–	57033.2	56846.1	56654.6	56458.8	56258.8	56054.4
40	–	76044.3	75794.8	75539.5	75278.4	75011.7	74739.3
50	–	95055.4	94743.4	94424.3	94098.0	93764.6	93424.1
60	–	114066.4	113692.1	113309.2	112917.7	112517.6	112108.9

Meridional Arcs—Values of D *m, in Yards.*

′	″		L. 21° 0′		L. 22° 0′		L. 23° 0′
	7		235.4		235.4		235.5
	8		269.0		269.1		269.1
	9		302.7		302.7		302.8
	10		336.3		336.4		336.4
	20		672.6		672.7		672.8
	30		1008.9		1009.1		1009.2
	40		1345.2		1345.4		1345.6
	50		1681.6		1681.8		1682.0
	60		2017.9		2018.1		2018.4
7	–		14125.0		14126.7		14128.5
8	–		16142.9		16144.8		16146.8
9	–		18160.8		18162.9		18165.2
10	–		20178.6		20181.0		20183.5
20	–		40357.2		40362.1		40367.1
30	–		60535.9		60543.1		60550.6
40	–		80714.5		80724.1		80734.1
50	–		100893.1		100905.2		100917.6
60	–		121071.7		121086.2		121101.2

Intermediate minutes and seconds will be found by moving the decimal point.

Arcs of the Parallel—Values of D p, in Yards.

$'$	$''$	L. 23° 30′	L. 24° 0′	L. 24° 30′	L. 25° 0′	L. 25° 30′	L. 26° 0′
	7	217.2	216.4	215.4	214.6	213.8	212.9
	8	248.2	247.3	246.3	245.3	244.3	243.3
	9	279.2	278.2	277.1	276.0	274.8	273.7
	10	310.3	309.1	307.9	306.6	305.4	304.1
	20	620.5	618.1	615.7	613.3	610.8	608.2
	30	930.8	927.6	923.6	919.9	916.2	912.3
	40	1241.0	1236.3	1231.5	1226.5	1221.5	1216.4
	50	1551.3	1545.4	1539.3	1533.2	1526.9	1520.5
	60	1861.5	1854.4	1847.2	1839.8	1832.3	1824.7
7	–	13030.7	12981.0	12930.4	12878.8	12826.1	12772.6
8	–	14892.2	14835.5	14777.6	14718.6	14658.5	14597.2
9	–	16753.8	16689.9	16624.8	16558.4	16490.8	16421.9
10	–	18615.3	18544.3	18472.0	18398.2	18323.1	18246.5
20	–	37230.6	37088.7	36944.0	36796.5	36646.1	36493.0
30	–	55845.8	55633.0	55416.0	55194.7	54969.2	54739.6
40	–	74461.1	74177.4	73887.9	73592.9	73292.3	72986.1
50	–	93076.4	92721.7	92359.9	91991.1	91615.3	91232.6
60	–	111691.7	111266.0	110831.9	110389.4	109938.4	109479.1

Meridional Arcs—Values of D m, in Yards.

$'$	$''$	L. 24° 0′	L. 25° 0′	L. 26° 0′
	7	235.5	235.5	235.6
	8	269.1	269.2	269.2
	9	302.8	302.8	302.9
	10	336.4	336.5	336.5
	20	672.9	673.0	673.1
	30	1009.3	1009.4	1009.6
	40	1345.7	1345.9	1346.1
	50	1682.2	1682.4	1682.6
	60	2018.6	2018.9	2019.2
7	–	14130.3	14132.1	14134.1
8	–	16148.9	16151.0	16153.2
9	–	18167.5	18169.9	18172.4
10	–	20186.1	20188.8	20191.5
20	–	40372.2	40377.5	40382.0
30	–	60558.3	60566.3	60574.6
40	–	80744.4	80755.1	80766.1
50	–	100930.5	100943.9	100957.6
60	–	121116.6	121132.6	121149.1

Intermediate minutes and seconds will be found by moving the decimal point.

Arcs of the Parallel—Values of D *p, in Yards.*

'	''	L. 26° 30'	L. 27° 0'	L. 27° 30'	L. 28° 0'	L. 28° 30'	L. 29° 0'
	7	212.0	211.0	210.1	209.1	208.2	207.2
	8	242.2	241.2	240.1	239.0	237.9	236.8
	9	272.5	271.3	270.1	268.9	267.6	266.4
	10	302.8	301.5	300.1	298.8	297.4	296.0
	20	605.6	603.0	600.3	597.6	594.8	591.9
	30	908.4	904.5	900.4	896.3	892.2	887.9
	40	1211.2	1206.0	1200.6	1195.1	1189.5	1183.9
	50	1514.0	1507.4	1500.7	1493.9	1486.9	1479.9
	60	1816.9	1808.9	1800.9	1792.7	1784.3	1775.8
7	–	12718.0	12662.5	12606.0	12548.6	12490.1	12430.8
8	–	14534.9	14471.4	14406.9	14341.2	14274.5	14206.6
9	–	16351.7	16280.3	16207.7	16133.9	16058.8	15982.5
10	–	18168.6	18089.3	18008.6	17926.5	17843.1	17758.3
20	–	36337.2	36178.5	36017.1	35853.0	35686.2	35516.6
30	–	54505.8	54267.8	54025.7	53779.5	53529.3	53274.9
40	–	72674.4	72357.1	72034.3	71706.1	71372.4	71033.2
50	–	90843.0	90446.3	90042.9	89632.6	89215.4	88791.5
60	–	109011.5	108535.6	108051.4	107559.1	107058.5	106549.8

Meridional Arcs—Values of D *m, in Yards.*

'	''	L. 27° 0'	L. 28° 0'	L. 29° 0'
	7	235.6	235.6	235.7
	8	269.3	269.3	269.3
	9	302.9	303.0	303.0
	10	336.6	336.6	336.7
	20	673.1	673.2	673.3
	30	1009.7	1009.9	1010.0
	40	1346.3	1346.5	1346.7
	50	1682.9	1683.1	1683.3
	60	2019.4	2019.7	2020.0
7	–	14136.0	14138.1	14140.1
8	–	16155.5	16157.8	16160.2
9	–	18174.9	18177.5	18180.2
10	–	20194.3	20197.2	20200.2
20	–	40388.7	40394.5	40400.4
30	–	60583.0	60591.7	60600.6
40	–	80777.4	80788.9	80800.8
50	–	100971.7	100986.2	101001.0
60	–	121166.0	121183.4	121201.2

Intermediate minutes and seconds will be found by moving the decimal point.

Arcs of the Parallel—Values of D p, in Yards.

′	″	L. 29° 30′	L. 30° 0′	L. 30° 30′	L. 31° 0′	L. 31° 30′	L. 32° 0′
	7	206.2	205.2	204.1	203.1	202.0	200.9
	8	235.6	234.5	233.3	232.1	230.9	229.6
	9	265.1	263.8	262.4	261.1	259.7	258.3
	10	294.5	293.1	291.6	290.1	288.6	287.0
	20	589.1	586.2	583.2	580.2	577.1	574.0
	30	883.6	879.2	874.8	870.3	865.7	861.1
	40	1178.1	1172.3	1166.4	1160.4	1154.3	1148.1
	50	1472.7	1465.4	1458.0	1450.5	1442.9	1435.1
	60	1767.2	1758.5	1749.6	1740.6	1731.4	1722.1
7	–	12370.5	12309.3	12247.1	12184.0	12120.0	12055.0
8	–	14137.7	14067.7	13996.7	13924.6	13851.4	13777.1
9	–	15904.9	15826.2	15746.3	15665.1	15582.8	15499.3
10	–	17672.2	17584.7	17495.9	17405.7	17314.2	17221.4
20	–	35344.3	35169.4	34991.7	34811.4	34628.4	34442.9
30	–	53016.5	52754.0	52487.6	52217.1	51942.7	51664.3
40	–	70688.7	70338.7	69983.4	69622.8	69256.9	68885.7
50	–	88360.8	87923.4	87479.3	87028.5	86571.1	86107.1
60	–	106033.0	105508.1	104975.2	104434.2	103885.3	103328.6

Meridional Arcs—Values of D m, in Yards.

′	″		L. 30° 0′		L. 31° 0′		L. 32° 0′
	7		235.7		235.7		235.8
	8		269.4		269.4		269.5
	9		303.0		303.1		303.1
	10		336.7		336.8		336.8
	20		673.4		673.5		673.6
	30		1010.2		1010.3		1010.5
	40		1346.9		1347.1		1347.3
	50		1683.6		1683.9		1684.1
	60		2020.3		2020.6		2020.9
7	–		14142.3		14144.4		14146.6
8	–		16162.6		16165.1		16167.6
9	–		18182.9		18185.7		18188.5
10	–		20203.2		20206.3		20209.5
20	–		40406.5		40412.6		40418.9
30	–		60609.7		60619.0		60628.4
40	–		80812.9		80825.3		80837.9
50	–		101016.1		101031.6		101047.4
60	–		121219.4		121237.9		121256.8

Intermediate minutes and seconds will be found by moving the decimal point.

Arcs of the parallel—Values of D p, in Yards.

′	″	L. 32° 30′	L. 33° 0″	L. 33° 30′	L. 34° 0′	L. 34° 30′	L.35° 0′
	7	199.8	198.7	197.6	196.4	195.3	194.1
	8	228.4	227.1	225.8	224.5	223.2	221.8
	9	256.9	254.5	254.0	252.6	251.1	249.6
	10	285.5	283.9	282.3	280.6	279.0	277.3
	20	570.9	567.7	564.5	561.2	557.9	554.6
	30	856.4	851.6	846.8	841.9	836.9	831.9
	40	1141.8	1135.5	1129.0	1122.5	1115.9	1109.2
	50	1427.3	1419.3	1411.3	1403.1	1394.8	1386.4
	60	1712.7	1703.2	1693.5	1683.7	1673.8	1663.7
7	–	11989.1	11922.3	11854.6	11786.0	11716.5	11646.1
8	–	13701.9	13625.5	13548.1	13469.7	13390.3	13309.8
9	–	15414.6	15328.7	15241.7	15153.5	15064.1	14973.6
10	–	17127.3	17031.9	16935.2	16837.2	16737.9	16637.3
20	–	34254.6	34063.8	33870.4	33674.3	33475.8	33274.6
30	–	51381.9	51095.7	50805.5	50511.5	50213.6	49911.9
40	–	68509.3	68127.6	67740.7	67348.7	66951.5	66549.2
50	–	85636.6	85159.5	84675.9	84185.8	83689.4	83186.5
60	–	102763.9	102191.4	101611.1	101023.0	100427.3	99823.8

Meridional Arcs—Values of D m, in Yards.

′	″	L. 33° 0′	L. 34° 0′	L. 35° 0′
	7	235.8	235.9	235.9
	8	269.5	269.5	269.6
	9	303.2	303.2	303.3
	10	336.9	336.9	337.0
	20	673.8	673.9	674.0
	30	1010.6	1010.8	1011.0
	40	1347.5	1347.7	1347.9
	50	1684.4	1684.7	1684.9
	60	2021.3	2021.6	2021.9
7	–	14148.9	14151.2	14153.5
8	–	16170.1	16172.7	16175.4
9	–	18191.4	18194.3	18197.3
10	–	20212.7	20215.9	20219.2
20	–	40425.4	40431.9	40438.5
30	–	60638.0	60647.8	60657.7
40	–	80850.7	80863.7	80877.0
50	–	101063.4	101079.7	101096.2
60	–	121276.1	121295.6	121315.4

Intermediate minntes and seconds will be found by moving the decimal point.

Arcs of the Parallel—Values of D *p, in Yards.*

′	″	L. 35° 30′	L. 36° 0′	L. 36° 30′	L. 37° 0′	L. 37° 30′	L. 38° 0′
	7	192.9	191.7	190.5	189.3	188.0	186.8
	8	220.5	219.1	217.7	216.3	214.9	213.4
	9	248.0	246.5	244.9	243.3	241.7	240.1
	10	275.6	273.9	272.1	270.4	268.6	266.8
	20	551.2	547.7	544.3	540.7	537.2	533.6
	30	826.8	821.6	816.4	811.1	805.8	800.4
	40	1102.4	1095.5	1088.5	1081.5	1074.4	1067.2
	50	1378.0	1369.4	1360.7	1351.9	1343.0	1334.0
	60	1653.5	1643.2	1632.9	1622.2	1611.6	1600.7
7	–	11574.8	11502.7	11429.6	11355.7	11280.9	11205.2
8	–	13228.4	13145.9	13062.4	12977.9	12892.5	12806.0
9	–	14881.9	14789.1	14695.2	14600.2	14504.0	14406.7
10	–	16535.5	16432.4	16328.0	16222.4	16115.6	16007.5
20	–	33070.9	32864.7	32656.0	32444.8	32231.1	32015.0
30	–	49606.4	49297.1	48984.0	48667.2	48346.7	48022.5
40	–	66141.9	65729.5	65312.1	64889.6	64462.3	64030.0
50	–	82677.3	82161.8	81640.1	81112.0	80577.8	80037.5
60	–	99212.8	98594.2	97968.1	97334.4	96693.4	96045.0

Meridional Arcs— Values of D *m, in Yards.*

′	″	Log 36° 0′	L. 37° 0′	L. 38° 0′
	7	235.9	236.0	236.0
	8	269.6	269.7	269.7
	9	303.3	303.4	303.4
	10	337.0	337.1	337.2
	20	674.1	674.2	674.3
	30	1011.1	1011.3	1011.5
	40	1348.2	1348.4	1348.6
	50	1685.2	1685.5	1685.8
	60	2022.3	2022.6	2022.9
7	–	14155.8	14158.2	14160.6
8	–	16178.1	16180.8	16183.5
9	–	18200.3	18203.4	18206.5
10	–	20222.6	20226.0	20229.4
20	–	40445.2	40451.9	40458.8
30	–	60667.5	60677.9	60688.2
40	–	80890.3	80903.9	80917.6
50	–	101112.9	101129.9	101147.0
60	–	121335.5	121355.8	121376.4

Intermediate minutes and seconds will be found by moving the decimal point.

Arcs of the Parallel—Values of D *p, in Yards.*

′	″	L. 38° 30′	L. 39° 0′	L. 39° 30′	L. 40° 0′	L. 40° 30	L. 41° 0′
	7	185.5	184.2	182.9	181.6	180.2	178.9
	8	212.0	210.5	209.0	207.5	206.0	204.4
	9	238.5	236.8	235.1	233.4	231.7	230.0
	10	265.0	263.1	261.3	259.4	257.5	255.6
	20	529.9	526.3	522.5	518.8	515.0	511.1
	30	794.9	789.4	783.8	778.2	772.4	766.7
	40	1059.9	1052.5	1045.1	1037.5	1029.9	1022.2
	50	1324.9	1315.6	1306.3	1296.9	1287.4	1277.8
	60	1589.8	1578.8	1567.6	1556.3	1544.9	1533.4
7	–	11128.7	11051.4	10973.2	10894.1	10814.3	10733.6
8	–	12718.6	12630.2	12540.8	12450.4	12359.2	12266.9
9	·	14308.4	14208.9	14108.4	14006.8	13904.1	13800.3
10	–	15898.2	15787.7	15676.0	15563.1	15448.9	15333.4
20	–	31796.4	31575.4	31351.9	31126.1	30897.9	30667.3
30	–	47694.6	47363.1	47027.9	46689.2	46346.8	46001.0
40	–	63592.6	63150.8	62703.9	62252.2	61795.8	61334.6
50	–	79491.0	78938.4	78379.9	77815.3	77244.7	76668.3
60	–	95389.2	94726.1	94055.8	93378.3	92693.7	92001.9

Meridional Arcs— Values of D *m, in Yards.*

′		L. 39° 0′		L. 40° 0′		L. 40° 0′
	7	236.0		236.1		236.1
	8	269.8		269.8		269.9
	9	303.5		303.5		303.6
	10	337.2		337.3		337.3
	20	674.4		674.5		674.7
	30	1011.6		1011.8		1012.0
	40	1348.9		1349.1		1349.3
	50	1686.1		1686.4		1686.7
	60	2023.3		2023.6		2024.0
7	–	14163.0		14165.4		14167.9
8	–	16186.3		16189.1		16191.9
9	–	18209.6		18212.7		18215.8
10	–	20232.8		20236.3		20239.8
20	–	40465.7		40472.7		40479.7
30	–	60698.5		60709.0		60719.5
40	–	80931.4		80945.3		80959.3
50	–	101164.2		101181.6		101199.2
60	–	121397.1		121418.0		121439.0

Intermediate minutes and seconds will be found by moving the decimal point.

Arcs of the Parallel—Values of D p, in Yards.

′	″	L. 41° 30′	L. 42° 0′	L. 42° 30′	L. 43° 0′	L. 43° 30′	L. 44° 0′
	7	177.5	176.2	174.8	173.4	172.0	170.5
	8	202.9	201.3	199.7	198.1	196.5	194.9
	9	228.3	226.5	224.7	222.9	221.1	219.3
	10	253.6	251.7	249.7	247.7	245.7	243.6
	20	507.2	503.3	449.4	495.4	491.3	487.3
	30	760.9	755.0	749.0	743.0	737.0	730.9
	40	1014.5	1006.6	998.7	990.7	982.7	974.5
	50	1268.1	1258.3	1248.4	1238.4	1228.3	1218.1
	60	1521.7	1510.0	1498.1	1486.1	1474.0	1461.8
7	–	10652.0	10569.7	10486.6	10402.6	10317.9	10232.3
8	–	12173.8	12079.7	11984.6	11888.7	11791.8	11694.1
9	–	13695.5	13589.6	13482.7	13374.8	13265.8	13155.8
10	–	15217.2	15099.6	14980.8	14860.9	14739.8	14617.6
20	–	30434.4	30199.1	29961.6	29721.7	29479.6	29235.2
30	–	45651.6	45298.7	44942.4	44582.6	44219.4	43852.8
40	–	60868.8	60398.3	59923.2	59443.4	58959.2	58470.4
50	–	76086.0	75497.9	74903.9	74304.3	73698.9	73088.0
60	–	91303.2	90597.4	89884.7	89165.1	88438.7	87705.6

Meridional Arcs—Values of D m, in Yards.

′	″	L. 42° 0′	L. 43° 0′	L. 44° 0′
	7	236.2	236.2	236.3
	8	269.9	270.0	270.0
	9	303.7	303.7	303.8
	10	337.4	337.4	337.5
	20	674.8	674.9	675.0
	30	1012.2	1012.3	1012.5
	40	1349.6	1349.8	1350.0
	50	1686.9	1687.2	1687.5
	60	2024.3	2024.7	2025.0
7	–	14170.3	14172.8	14175.3
8	–	16194.7	16197.5	16200.3
9	–	18219.0	18222.2	18225.4
10	–	20243.4	20246.9	20250.4
20	–	40486.7	40493.8	40500.9
30	–	60730.1	60740.7	60751.3
40	–	80973.4	80987.5	81001.7
50	–	101216.8	101234.4	101252.2
60	–	121460.1	121481.3	121502.6

Intermediate minutes and seconds will be found by moving the decimal point.

Arcs of the Parallel—Values of D *p, in Yards.*

′	″	L. 44° 30′	L. 45° 0′	L. 45° 30′	L. 46° 0′	L. 46° 30′	L. 47° 0′
	7	169.1	167.6	166.2	164.7	163.2	161.7
	8	193.3	191.6	189.9	188.2	186.5	184.8
	9	217.4	215.5	213.7	211.8	209.8	207.9
	10	241.6	239.5	237.4	235.3	233.2	331.0
	20	483.1	479.0	474.8	470.6	466.3	462.0
	30	724.7	718.5	712.2	705.9	699.5	693.1
	40	966.3	958.0	949.6	941.2	932.7	924.1
	50	1207.9	1197.5	1187.0	1176.5	1165.8	1155.1
	60	1449.4	1437.0	1424.4	1411.8	1399.0	1386.1
7	–	10146.0	10058.9	9971.0	9882.4	9793.0	9702.8
8	–	11595.4	11495.9	11395.5	11294.2	11192.0	11089.0
9	–	13044.8	12932.9	12819.9	12705.9	12591.0	12475.1
10	–	14494.3	14369.8	14244.3	14117.7	13990.0	13861.2
20	–	28988.6	28739.7	28488.6	28235.4	27980.0	27722.4
30	–	43482.8	43109.5	42733.0	42353.1	41970.0	41583.6
40	–	57977.1	57479.4	56977.3	56470.8	55960.0	55444.8
50	–	72471.4	71849.2	71221.6	70588.5	69950.0	69306.0
60	–	86965.7	86219.1	85465.9	84706.2	83940.0	83167.3

Meridional Arcs—Values of D *m, in Yards.*

′	″	L. 45° 0′	L. 46° 0′	L. 47° 0′
	7	236.3	236.3	236.4
	8	270.1	270.1	270.1
	9	303.8	303.9	303.9
	10	337.6	337.6	337.7
	20	675.1	675.3	675.4
	30	1012.7	1012.9	1013.1
	40	1350.3	1350.5	1350.7
	50	1687.8	1688.1	1688.4
	60	2025.4	2025.8	2026.1
7	–	14177.8	14180.3	14182.8
8	–	16203.2	16206.0	16208.9
9	–	18228.6	18231.8	18235.0
10	–	20254.0	20257.5	20261.1
20	–	40508.0	40515.1	40522.2
30	–	60761.9	60772.6	60783.2
40	–	81015.9	81030.1	81044.3
50	–	101269.9	101287.7	101305.4
60	–	121523.9	121545.2	121566.5

Intermediate minutes and seconds will be found by moving the decimal point.

Arcs of the Parallel—Values of D p, in Yards.

$'$	$''$	L. 47° 30′	L. 48° 0′	L. 48° 30′	L. 49° 0′	L. 49° 30′	L. 50° 0′
	7	160.2	158.7	157.1	155.6	154.0	152.4
	8	183.1	181.3	179.6	177.8	176.0	174.2
	9	206.0	204.0	202.0	200.0	198.0	196.0
	10	228.9	226.7	224.5	222.3	220.0	217.8
	20	457.7	453.3	449.0	444.5	440.1	435.6
	30	686.6	680.0	673.4	666.8	660.1	653.3
	40	915.4	906.7	897.9	889.0	880.1	871.1
	50	1144.3	1133.4	1122.4	1111.3	1100.1	1088.9
	60	1373.1	1360.0	1346.9	1333.6	1320.2	1306.7
7	–	9612.0	9520.3	9428.0	9334.9	9241.1	9146.6
8	–	10985.1	10880.4	10774.8	10668.5	10561.2	10453.2
9	–	12358.2	12240.4	12121.7	12002.0	11881.4	11759.9
10	–	13731.4	13600.5	13468.5	13335.6	13201.6	13066.5
20	–	27462.7	27200.9	26937.1	26671.1	26403.1	26133.1
30	–	41194.1	40801.4	40405.6	40006.7	39604.7	39199.6
40	–	54925.5	54401.9	53874.1	53342.3	52806.2	52266.2
50	–	68656.8	68002.4	67342.7	66677.8	66007.8	65332.7
60	–	82388.2	81602.8	80811.2	80013.4	79209.4	78399.3

Meridional Arcs—Values of D m, in Yards.

$'$	$''$	L. 48° 0′	L. 49° 0′	L. 50° 0′
	7	236.4	236.5	236.5
	8	270.2	270.2	270.3
	9	304.0	304.0	304.1
	10	337.7	337.8	337.9
	20	675.5	675.6	675.7
	30	1013.2	1013.4	1013.6
	40	1351.0	1351.2	1351.4
	50	1688.7	1689.0	1689.3
	60	2026.5	2026.8	2027.2
7	–	14185.2	14187.7	14190.2
8	–	16211.7	16214.5	16217.3
9	–	18238.2	18241.3	18244.5
10	–	20264.6	20268.1	20271.7
20	–	40529.2	40536.3	40543.3
30	–	60793.9	60804.4	60815.0
40	–	81058.5	81072.6	81086.6
50	–	101323.1	101340.7	101358.9
60	–	121587.7	121608.9	121629.9

Intermediate minutes and seconds will be found by moving the decimal point.

Lengths in Nautical miles and Statute miles of degrees of Latitude and Longitude in different Latitudes.

DEGREE OF THE PARALLEL.			DEGREE OF THE MERIDIAN.		
Latitude of Parallel.	Nautical miles.	Statute miles.	Latitude of middle point.	Nautical miles.	Statute miles.
20°	56.404	65.018	20°	59.664	68.777
21	56.039	64.598			
22	55.657	64.158			
23	55.258	63.698			
24	54.843	63.219			
25	54.411	62.721	25	59.706	68.825
26	53.962	62.204			
27	53.497	61.668			
28	53.016	61.113			
29	52.518	60.540			
30	52.005	59.948	30	59.749	68.875
31	51.476	59.338			
32	50.931	58.709			
33	50.370	58.063			
34	49.794	57.399			
35	49.203	56.718	35	59.796	68.929
36	48.597	56.019			
37	47.976	55.304			
38	47.341	54.571			
39	46.960	53.822			
40	46.026	53.056	40	59.847	68.987
41	45.348	52.274			
42	44.654	51.476			
43	43.949	50.662			
44	43.230	49.833			
45	42.497	48.988	45	59.899	69.048
46	41.752	48.128			
47	40.993	47.254			
48	40.222	46.365			
49	39.439	45.462			
50	38.643	44.545	50	59.951	69.108

A degree of longitude at the equator = 69.163 statute miles.
A second of time at the equator = 1521.6 feet.

APPENDIX.

MAGNETIC OBSERVATIONS.

On the use of the Portable Declinometer in the determination of the magnetic Declination (variation) and Horizontal Intensity.

ABSOLUTE DECLINATION.

The adjustment of the Declinometer consists in bringing the line of detorsion (or the direction in which the force of torsion of the thread acts,) first approximately, then accurately, into the magnetic meridian; in determining the zero division of the scale corresponding to the magnetic axis of the collimator magnet, and in bringing the line of collimation of the Theodolite telescope into the magnetic meridian, its vertical wire coinciding with this division.

Having determined the readings of the verniers of the Azimuth circle of the Theodolite corresponding to the magnetic axis of the collimator magnet, turn the telescope into the direction of some object whose azimuth from true north is known, or can be determined, and read off again. The difference of these readings added to or subtracted from the true azimuth of the object referred to, will give the absolute declination.

As the direction of the magnetic meridian is continually changing, the instrument should be left in adjustment, and observed half-hourly, or hourly, for as long a period as possible, in order that the mean declination may be obtained,

17

instead of the declination at one period only.   By this means, also, variations of declination will be obtained.

The angular value of the magnet's scale is determined by measuring with the theodolite the horizontal angle subtended by a certain number of its divisions, the magnet being temporarily fixed.   As the interval of the lens and scale is adjusted by the maker, so that the divisions shall be most clearly seen at infini-distant focus, if the adjustment is very accurately made the angular value of the scale will be the same at whatever distance the telescope of the Theodolite may be placed.   This, however, should always be tested.

If a, denote the angular value of one division of the scale, and $\dfrac{H}{F}$ the ratio of the torsion and magnetic forces, the true declination changes are deduced from the observed readings, by multiplying their differences by the constant co-efficient $a \times \left(1 + \dfrac{H}{F} \right)$

The value of $\dfrac{H}{F}$ is determined by turning the torsion circle through two or more large angles (for example 90°) and noting the corresponding differences of reading; if w equals the mean of the former, and u that of the latter, reduced to angular value, $\dfrac{H}{F} = \dfrac{w}{w - u}$.

The value of the co-efficient $a \left(1 + \dfrac{H}{F} \right)$ must be given with every abstract of observations; and also a statement of whether increasing numbers denote an easterly or westerly movement of the north end of the magnet.

ABSOLUTE HORIZONTAL INTENSITY.

The determination of the horizontal intensity requires two distinct series of observations or experiments; those of deflection and those of vibration.

Experiments of Deflection give the ratio of the magnetic moment of the deflecting magnet to the Horizontal Intensity.

Experiments of Vibration, the product of the same quantities. The absolute value of either is obtained by comparing the two.

The experiments of deflection consist in observing the angles through which a freely suspended magnet is deflected, by the action of a second magnet, placed at different distances from it in the direction of a line passing through its centre, perpendicular either to the axis of the suspended magnet, or to the magnetic meridian.

The experiments of deflection with the portable declinometer should be made, if possible, at three distances at each side of the suspended magnet, reversing the deflecting magnet each time, and repeating the reversals so as to obtain four readings at each distance on each side. The first distance should be as near as the length of the collimator scale will allow. The second distance, if possible, one-third greater than the first, and the third intermediate to the two first.

When the deflecting magnet is shifted from the east to the west side of the suspended magnet, reverse the order of distances, so that the mean results may correspond nearly to the same time of observation. The corrections for changes of temperature and intensity (during the time employed in the observations) will, in the absence of proper instruments for observing them, in most cases, be compensated by this system of observation.

Calculation.—The ratio of the magnetic moment of the deflecting magnet to the horizontal intensity is to be calculated in any case from the observations at each distance taken separately ; if

$m =$ the magnetic moment of the deflecting magnet,

$X =$ the horizontal intensity,

$r =$ the distance between the centres of the deflecting and suspended magnets expressed in feet and decimals of a foot,

$u =$ the corresponding angle of deflection, obtained by multiplying half the mean of each partial result (at each distance) by the angular value of one scale division corrected for torsion,

then: $$\frac{m}{X} = \tfrac{1}{2}\, r^3 \, \text{tang } u .$$

The experiments of vibration consist in suspending the magnet, used as a deflecting bar, in a wooden box of suitable dimensions, and noting the times at which some central division of the scale passes across the vertical wire of the telescope (of the Theodolite) during at least 300 vibrations ; the magnet having been first made to vibrate steadily, and the arc of vibration reduced to as small a size as the observations will allow.

As the time of vibration depends on the form and weight of the suspended mass, as well as on the product of the magnetic moment of the magnet into the horizontal intensity, its moment of inertia must be experimentally determined before the required value of this product can be obtained. The experiment consists in observing a second series of vibrations with two cylindrical weights of equal dimensions, whose moment of inertia is known, suspended at opposite ends of the magnet; if,

$K =$ the moment of the suspended magnet and stirrup, (the value required,)
$K' =$ the moment of inertia of the weights,
T' and $T =$ the times of vibration with and without the weights,

$$K = K' \left(\frac{T'}{T'^2 - T^2} \right)$$

the moment of inertia of the weights is calculated by the formula,

$$K' = \left(\tfrac{1}{2} l^2 + r^2 \right) p$$

where :

$l =$ interval of the points of suspension, or length of the magnet diminished by the depth of the grooves in which the threads rest, $r =$ the radius of the cylinders, and $2\,p$ their mass, expressed respectively in feet and grains.

The value of K must be ascertained with very great accuracy from the mean of several series of observations with and without the weights; once satisfactorily determined, it may be employed as a constant quantity, and the observations with weights need not be repeated in after determinations of the intensity. A small correction must be applied for the changes in the dimensions or form of the suspended mass, produced by changes of temperature; the correction consists in multiplying the value of K by $1 + 2e(t' - t)$, where t' denotes the actual temperature of the magnet, t the temperature corresponding to the time of the original observations, and e the coefficient of dilatation of steel for 1° Fahrenheit $= 0.0000068$.

Calculation.—The product of the horizontal intensity into the magnetic moment of the suspended magnet is obtained by the formula :

$$m X = \frac{\pi^2 K}{T^2}$$

where $\pi =$ circumference of circle to diameter 1.

     $K =$ the moment of inertia of suspended magnet and stirrup.

     $T =$ the time of vibration, corrected.

The observed times of vibration must be corrected for the force of torsion of the suspension thread, the rate of chronometer, the arc of vibration, the change of horizontal force between the observations of deflection and vibration, and the differences in the magnetic moment of the deflecting magnet, produced by an increase or decrease of temperature, and by the earth's inducing action. The corrections are to be applied according to the formula :

$$T^2 = \left\{ T'\left(1 - \frac{r}{86400}\right) \cdot \left(1 - \frac{a\,a'}{16}\right) \right\}^2 \cdot \left(1 + \frac{H}{F}\right) \cdot$$
$$\left(1 + \frac{\Delta X}{X}\right) \cdot \left(1 - (t' - t)\,q\right) \cdot \left(1 + \frac{\Delta m}{m}\right) \cdot$$

$\frac{\Delta X}{X}$ the change of horizontal force between the experiments of deflection and vibration, and $\frac{\Delta m}{m}$ the difference in the magnetic moment of the deflecting magnet, produced by the earth's inducing action, are determined from experiments with the Bifilar magnetometer—when this instrument is not used, as is the case in our observations and upon magnetic surveys generally, the formula becomes:

$$T^2 = \left\{ T' \left(1 - \frac{r}{86400} - \frac{a\,a'}{16} \right) \right\}^2 \times \left(1 + \frac{H}{F} \right)$$
$$\times \left(1 - (t' - t)\, q \right)$$

where:

T and T$'$ = the true and observed times of vibration,

r = the rate of the chronometer $\left\{ \begin{array}{l} - \text{when losing,} \\ + \text{when gaining} \end{array} \right.$

a and a' = the initial and terminal semi-arcs of vibration in parts of radius,

$\dfrac{H}{F}$ = the ratio of the torsion and magnetic forces,

t = the temperature of the deflecting magnet during the experiments of deflection,

t' = the temperature of the deflecting magnet during the experiments of vibration,

q = the change of magnetic moment of the deflecting magnet for 1° of temperature.

The value of $\dfrac{a\,a'}{16}$ is found by the formula:

$$\frac{a\,a'}{16} = a^2\, d\, d' \times 0.00007272^2$$

where d, d' denote the semi-arcs of vibration in divisions of scale, and a, the angular value of 1 division.*

$$q = \frac{1}{t - t_0}\, a \,.\, n \,.\, \cot u \,.$$

where $\quad a =$ the arc value of 1 division of scale in terms of radius,

$\quad\quad h =$ the difference of scale readings corrected for changes of declination,

$\quad\quad t - t_0 =$ the corresponding difference of temperature,

$\quad\quad u =$ the angle of deflection at the lowest mean temperature.

Final calculation of the results.

By the observations of deflection, we have $\dfrac{m}{X} = A$

By those of vibration, $\quad.\quad\quad.\quad\quad.\quad m\,X = B$

$$\text{whence,}\quad X = \sqrt{\frac{B}{A}}$$

$$\text{and}\quad\quad m = \sqrt{A\,B.}$$

---

*The correction for *arc*, in an extreme case when the initial semi-arc was 36 scale divisions ($1°\ 40'\ 30''$), amounted to only 0.000037.

The correction for *rate*, when the chronometer loses 3 seconds per day, is 0.000084.

From examples in Riddell's Instructions, the value of the horizontal intensity is not carried beyond four decimal figures, and as these two corrections only change the fifth decimal, it will be, in most instances, needless to compute them, and the final formula would then be further reduced to

$$T^2 = T'^2 \times \left(1 + \frac{H}{F}\right)\ \left(1 - (t' - t)\,q\right)\,.$$

The horizontal intensity being thus determined, the *Total Intensity* will be found by dividing the horizontal intensity by the cosine of the *Dip*, deduced from observations with the dipping needle.

OBSERVATIONS FOR THE DIP, OR ABSOLUTE INCLINATION.

In addition to the usual method of conducting observations of the Dip Circle, it is desirable that a few series should be made in different azimuths, for the purpose of testing the axles of the needles and the limb of the instrument (which is often magnetic ;) if η denote the observed inclination of the needle, θ the inclination sought, a the azimuth of the vertical circle.

$$\text{tang } \theta = \text{tang } \eta \text{ co-sec } a$$

The true inclination may be inferred also from the observed inclinations of the needle in any two planes at right angles to one another, without the knowledge of the angle a, according to the formula

$$\cot^2 \theta = \cot^2 \eta + \cot^2 \eta'$$

The difference between the mean of the results obtained by observations in different azimuths, and the result obtained by observations in the magnetic meridian, should be applied as a constant correction for the errors of axl^c and limb to all preceding or subsequent observations made in the meridian.

18

To compute, theoretically, the variations in the magnetic declination, due to changes in the Latitude and Longitude.

In a system of rectangular co-ordinates of which z_0 is the origin, let—

$z =$ declination at P.
$M =$ increase of declination in the direction, x.
$N =$ increase of declination in the direction, y,
$x =$ difference of longitude.
$y =$ difference of latitude.

At the origin,

$$\Sigma\, x = o\,; \qquad\qquad \Sigma\, y = o\,;$$

$$n\, L = \Sigma\, z\,; \qquad\qquad L = \frac{\Sigma\, z}{n}$$

$$M\, \Sigma\, x^2 + N\, \Sigma\, x\, y = \Sigma\, x\, z.$$

$$M\, \Sigma\, x\, y + N\, \Sigma\, y^2 = \Sigma\, y\, z.$$

$$\text{Let } z - L = K\,;\ \text{then,}$$

Equation of the line passing through all the points of equal declination is, $M\, x + N\, y = K$, and the angle which it makes with the meridian.

$$= \mathrm{arc}\left(\tan\! g = -\frac{N}{M}\right)$$

TABLES AND FORMULÆ.

PART III.

ASTRONOMY.

I. *Of Sidereal and Solar Time.*

True or *apparent Solar time* is that deduced from observations of the Sun, or is the same as that shown by a well adjusted sun-dial.

Mean Solar Time is derived from the time employed by the Earth in revolving on its axis, as compared with the Sun, supposed to move at a mean rate in its orbit, and to make 365.242218 revolutions in a mean Gregorian year. It cannot be immediately obtained from observation, but is always deduced from apparent time by the aid of the equation of time, which is the angular distance, in time, between the mean and true sun; or mean solar time = apparent solar time $\pm$ equation of time.

Sidereal Time is the portion of a sidereal day which has elapsed since the transit of the first point of Aries. Its point of origin cannot be determined by observation, but it is known at any moment by the right ascension of whatever object may be then in the meridian, or

Sidereal time of a star's culmination = AR of $\ast$

Sidereal time at mean moon = AR mean $\odot$ at mean moon; and, generally,

Sidereal time = sidereal time at mean moon $\pm$ solar time from mean moon (expressed in sidereal intervals.)

Solar time = sidereal time — sidereal time at mean moon, (the difference being reduced to a solar interval.)

Example—

To find the mean solar time of the passage of Altair over the meridian of Washington, on the 10th July, 1849.

		h	m	s
AR: Altair, July 10th, 1849 . . . . . =		19	43	27 39
Sidereal time at mean moon at Washington . . =		7	14	00.96
Sidereal interval past Washington mean moon . . =		12	29	26.43
Retardation of mean on sidereal time . . . = —			2	02.27
Corresponding mean time interval past mean moon or mean time of culmination . . . . . =		12	27	23.66

The nautical almanacs give the sidereal time at mean moon for each day of the year for a certain meridian.

If the sidereal day be taken equal to 24 sidereal hours, the mean solar day will be equal to 24^h 03^m $56^s.55$ of those sidereal hours; or the daily acceleration of sidereal on mean solar time (which is the mean motion of the earth in a mean solar day,) is 3^m $56^s.5554$ of sidereal time; hence the sidereal time at mean moon under any meridian other than that of the nautical almanac used, will be found by allowing the proportion of this quantity due to the difference of longitude of the two places.

If the mean solar day be taken equal to 24 mean solar hours, the sidereal day will be equal to 23^h 56^m $04^s.09$ of those solar hours, or the daily retardation of mean solar on sidereal time is 3^m $55^s.9093$ of solar time.

The astronomical day begins at noon. In the civil or common method of reckoning, the day is supposed to commence at the *preceding* midnight. The civil reckoning is therefore 12 hours in advance of the astronomical reckoning, and in the above example, July 10th, 12^h 27^m $23^s.66$ astronomical time, corresponds to July 11th, 0^h 27^m $23^s.66$ A. M. civil time.

II. *To find the time by an altitude of the Sun or a star.*

$$\text{Sidereal time} = AR \, \ast \pm \ast\text{'s hour angle,}$$
$$\text{Solar time} = 24^{hrs} \pm \odot\text{'s hour angle,}$$
$$2m = L + \Delta + A,$$
$$\text{Sin}^2 \tfrac{1}{2} p = \frac{\cos m \,.\, \sin (m-A)}{\cos L \,.\, \sin \Delta}, \text{ where}$$

$L =$ the latitude of the place of observation.

$\Delta =$ the north polar distance of the sun or the star.

$A =$ the corrected altitude of the sun or star.

$\quad =$ obs. alt. —(refrac'n—paral'x) $\pm$ semi-diam.

$p =$ the hour angle of the sun or star.

The formula gives the arc in degrees, which must be converted into time, as in one of the following four cases:

1. When we have the *corrected* altitude of the sun's *centre*, the hour angle, p, in time, is the *apparent* time when the sun is in the west, or the complement of 24 hours when in the east. To reduce it to *mean* time apply the equation of time.

2. But should the *sidereal* time be required, transform the mean time thus obtained to sidereal time, as previously explained.

3. When the altitude is that of a star, the sideral time is at once deduced from the hour angle p.

4. And if, in this last instance, solar time should be required, convert this sidereal time into solar time by means of the equation,

$$\text{Solar time} = AR \, \ast - AR \, \odot \pm p.$$

In which the sign $+$ is used if the star is observed in the west, and the sign $-$ if in the east; or,

Mean solar time $=$ the mean solar equivalent of (sid. time of obs'n—sid'l time of preced'g mean moon at place.)

For converting intervals of SIDEREAL *into corresponding intervals of* MEAN SOLAR *time.*

HOURS.			MINUTES.				SECONDS.			
h.	m.	s.	m.	s.	m.	s.	s.	s.	s.	s.
1	0	9.830	1	0.164	31	5.079	1	0.003	31	0.085
2	0	19.659	2	0.328	32	5.242	2	0.005	32	0.087
3	0	29.489	3	0.491	33	5.406	3	0.008	33	0.090
4	0	39.318	4	0.655	34	5.570	4	0.011	34	0.093
5	0	49.148	5	0.819	35	5.734	5	0.014	35	0.096
6	0	58.977	6	0.983	36	5.898	6	0.016	36	0.098
7	1	8.807	7	1.147	37	6.062	7	0.019	37	0.101
8	1	18.636	8	1.311	38	6.225	8	0.022	38	0.104
9	1	28.466	9	1.474	39	6.389	9	0.025	39	0.106
10	1	38.296	10	1.638	40	6.553	10	0.027	40	0.109
11	1	48.125	11	1.802	41	6.717	11	0.030	41	0.112
12	1	57.955	12	1.966	42	6.881	12	0.033	42	0.115
13	2	7.784	13	2.130	43	7.044	13	0.036	43	0.118
14	2	17.614	14	2.294	44	7.208	14	0.038	44	0.120
15	2	27.443	15	2.457	45	7.372	15	0.041	45	0.123
16	2	37.273	16	2.621	46	7.536	16	0.044	46	0.126
17	2	47.103	17	2.785	47	7.700	17	0.047	47	0.128
18	2	56.932	18	2.949	48	7.864	18	0.049	48	0.131
19	3	6.762	19	3.113	49	8.027	19	0.052	49	0.134
20	3	16.591	20	3.277	50	8.191	20	0.055	50	0.137
21	3	26.421	21	3.440	51	8.355	21	0.057	51	0.140
22	3	36.250	22	3.604	52	8.519	22	0.060	52	0.142
23	3	46.080	23	3.768	53	8.683	23	0.063	53	0.145
24	3	55.909	24	3.932	54	8.847	24	0.066	54	0.148
			25	4.096	55	9.010	25	0.068	55	0.150
			26	4.259	56	9.174	26	0.071	56	0.153
			27	4.423	57	9.338	27	0.074	57	0.156
			28	4.587	58	9.502	28	0.076	58	0.159
			29	4.751	59	9.666	29	0.079	59	0.161
			30	4.915	60	9.830	30	0.082	60	0.164

The quantities taken from this table must be *subtracted* from a sidereal interval, to obtain the corresponding interval in mean solar time.

For converting intervals of MEAN SOLAR *Time into corresponding intervals of* SIDEREAL *Time.*

HOURS.			MINUTES.				SECONDS.			
h.	m.	s.	m.	s.	m.	s.	s.	s.	s.	s.
1	0	9.856	1	0.164	31	5.092	1	0.003	31	0.085
2	0	19.713	2	0.329	32	5.257	2	0.005	32	0.088
3	0	29.569	3	0.493	33	5.421	3	0.008	33	0.090
4	0	39.426	4	0.657	34	5.585	4	0.011	34	0.093
5	0	49.282	5	0.821	35	5.750	5	0.014	35	0.096
6	0	59.139	6	0.986	36	5.914	6	0.016	36	0.098
7	1	8.995	7	1.150	37	6.078	7	0.019	37	0.101
8	1	18.852	8	1.314	38	6.242	8	0.022	38	0.104
9	1	28.708	9	1.478	39	6.407	9	0.025	39	0.106
10	1	38.565	10	1.643	40	6.571	10	0.027	40	0.109
11	1	48.421	11	1.807	41	6.735	11	0.030	41	0.112
12	1	58.278	12	1.971	42	6.900	12	0.033	42	0.115
13	2	8.134	13	2.136	43	7.064	13	0.036	43	0.118
14	2	17.991	14	2.300	44	7.228	14	0.038	44	0.120
15	2	27.847	15	2.464	45	7.392	15	0.041	45	0.123
16	2	37.704	16	2.628	46	7.557	16	0.044	46	0.126
17	2	47.560	17	2.793	47	7.721	17	0.047	47	0.128
18	2	57.416	18	2.957	48	7.885	18	0.049	48	0.131
19	3	7.273	19	3.121	49	8.050	19	0.052	49	0.134
20	3	17.129	20	3.285	50	8.214	20	0.055	50	0.137
21	3	26.986	21	3.450	51	8.378	21	0.057	51	0.140
22	3	36.842	22	3.614	52	8.542	22	0.060	52	0.142
23	3	46.699	23	3.778	53	8.707	23	0.063	53	0.145
24	4	56.555	24	3.943	54	8.871	24	0.066	54	0.148
			25	4.107	55	9.035	25	0.068	55	0.151
			26	4.271	56	9.199	26	0.071	56	0.153
			27	4.436	57	9.364	27	0.074	57	0.156
			28	4.600	58	9.528	28	0.077	58	0.159
			29	4.764	59	9.692	29	0.079	59	0.161
			30	4.928	60	9.856	30	0.082	60	0.164

The quantities taken from this table must be *added* to a mean interval, to obtain the corresponding interval in sidereal time.

To convert parts of the Equator in Arc into Sidereal Time, or to convert Terrestrial Longitude in Arc into Time.

DEGREES.

Arc.	Time.	Arc.	Time.	Arc.	Time.	Arc.	Time.
°	h m	°	h m	°	h m	°	h m
1	0 4	31	2 4	61	4 4	91	6 4
2	0 8	32	2 8	62	4 8	92	6 8
3	0 12	33	2 12	63	4 12	93	6 12
4	0 16	34	2 16	64	4 16	94	6 16
5	0 20	35	2 20	65	4 20	95	6 20
6	0 24	36	2 24	66	4 24	96	6 24
7	0 28	37	2 28	67	4 28	97	6 28
8	0 32	38	2 32	68	4 32	98	6 32
9	0 36	39	2 36	69	4 36	99	6 36
10	0 40	40	2 40	70	4 40	100	6 40
11	0 44	41	2 44	71	4 44	101	6 44
12	0 48	42	2 48	72	4 48	102	6 48
13	0 52	43	2 52	73	4 52	103	6 52
14	0 56	44	2 56	74	4 56	104	6 56
15	1 0	45	3 0	75	5 0	105	7 0
16	1 4	46	3 4	76	5 4	106	7 4
17	1 8	47	3 8	77	5 8	107	7 8
18	1 12	48	3 12	78	5 12	108	7 12
19	1 16	49	3 16	79	5 16	109	7 16
20	1 20	50	3 20	80	5 20	110	7 20
21	1 24	51	3 24	81	5 24	111	7 24
22	1 28	52	3 28	82	5 28	112	7 28
23	1 32	53	3 32	83	5 32	113	7 32
24	1 36	54	3 36	84	5 36	114	7 36
25	1 40	55	3 40	85	5 40	115	7 40
26	1 44	56	3 44	86	5 44	116	7 44
27	1 48	57	3 48	87	5 48	117	7 48
28	1 52	58	3 52	88	5 52	118	7 52
29	1 56	59	3 56	89	5 56	119	7 56
30	2 0	60	4 0	90	6 0	120	8 0

To convert parts of the Equator in Arc into Sidereal Time, or to convert Terrestrial Longitude in Arc into Time.

DEGREES.

Arc.	Time.		Arc.	Time.		Arc.	Time.		Arc.	Time.	
°	h	m	°	h	m	°	h	m	°	h	m
121	8	4	151	10	4	181	12	4	211	14	4
122	8	8	152	10	8	182	12	8	212	14	8
123	8	12	153	10	12	183	12	12	213	14	12
124	8	16	154	10	16	184	12	16	214	14	16
125	8	20	155	10	20	185	12	20	215	14	20
126	8	24	156	10	24	186	12	24	216	14	24
127	8	28	157	10	28	187	12	28	217	14	28
128	8	32	158	10	32	188	12	32	218	14	32
129	8	36	159	10	36	189	12	36	219	14	36
130	8	40	160	10	40	190	12	40	220	14	40
131	8	44	161	10	44	191	12	44	221	14	44
132	8	48	162	10	48	192	12	48	222	14	48
133	8	52	163	10	52	193	12	52	223	14	52
134	8	56	164	10	56	194	12	56	224	14	56
135	9	0	165	11	0	195	13	0	225	15	0
136	9	4	166	11	4	196	13	4	226	15	4
137	9	8	167	11	8	197	13	8	227	15	8
138	9	12	168	11	12	198	13	12	228	15	12
139	9	16	169	11	16	199	13	16	229	15	16
140	9	20	170	11	20	200	13	20	230	15	20
141	9	24	171	11	24	201	13	24	231	15	24
142	9	28	172	11	28	202	13	28	232	15	28
143	9	32	173	11	32	203	13	32	233	15	32
144	9	36	174	11	36	204	13	36	234	15	36
145	9	40	175	11	40	205	13	40	235	15	40
146	9	44	176	11	44	206	13	44	236	15	44
147	9	48	177	11	48	207	13	48	237	15	48
148	9	52	178	11	52	208	13	52	238	15	52
149	9	56	179	11	56	209	13	56	239	15	56
150	10	0	180	12	0	210	14	0	240	16	0

To convert parts of the Equator in Arc into Sidereal Time, or to convert Terrestrial Longitude in Arc into Time.

DEGREES.

Arc.	Time.	Arc.	Time.	Arc.	Time.	Arc.	Time.
°	h m	°	h m	°	h m	°	h m
241	16 4	271	18 4	301	20 4	331	22 4
242	16 8	272	18 8	302	20 8	332	22 8
243	16 12	273	18 12	303	20 12	333	22 12
244	16 16	274	18 16	304	20 16	334	22 16
245	16 20	275	18 20	305	20 20	335	22 20
216	16 24	276	18 24	306	20 24	336	22 24
247	16 28	277	18 28	307	20 28	337	22 28
248	16 32	278	18 32	308	20 32	338	22 32
249	16 36	279	18 36	309	20 36	339	22 36
250	16 40	280	18 40	310	20 40	340	22 40
251	16 44	281	18 44	311	20 44	341	22 44
252	16 48	282	18 48	312	20 48	342	22 48
253	16 52	283	18 52	313	20 52	343	22 52
254	16 56	284	18 56	314	20 56	344	22 56
255	17 0	285	19 0	315	21 0	345	23 0
256	17 4	286	19 4	316	21 4	346	23 4
257	17 8	287	19 8	317	21 8	347	23 8
258	17 12	288	19 12	318	21 12	348	23 12
259	17 16	289	19 16	319	21 16	349	23 16
260	17 20	290	19 20	320	21 20	350	23 20
261	17 24	291	19 24	321	21 24	351	23 24
262	17 28	292	19 28	322	21 28	352	23 28
263	17 32	293	19 32	323	21 32	353	23 32
264	17 36	294	19 36	324	21 36	354	23 36
265	17 40	295	19 40	325	21 40	355	23 40
266	17 44	296	19 44	326	21 44	356	23 44
267	17 48	297	19 48	327	21 48	357	23 48
268	17 52	298	19 52	328	21 52	358	23 52
269	17 56	299	19 56	329	21 56	359	23 56
270	18 0	300	20 0	330	22 0	360	24 0

To convert parts of the Equator in Arc into Sidereal Time or to convert Terrestrial Longitude in Arc into Time.

MINUTES.				SECONDS.			
Arc.	Time.	Arc.	Time.	Arc.	Time.	Arc.	Time.
′	m s	′	m s	″	s	″	s
1	0 4	31	2 4	1	0.067	31	2.067
2	0 8	32	2 8	2	0.133	32	2.133
3	0 12	33	2 12	3	0.200	33	2.200
4	0 16	34	2 16	4	0.267	34	2.267
5	0 20	35	2 20	5	0.333	35	2.333
6	0 24	36	2 24	6	0.400	36	2.400
7	0 28	37	2 28	7	0.467	37	2.467
8	0 32	38	2 32	8	0.533	38	2.533
9	0 36	39	2 36	9	0.600	39	2.600
10	0 40	40	2 40	10	0.667	40	2.667
11	0 44	41	2 44	11	0.733	41	2.733
12	0 48	42	2 48	12	0.800	42	2.800
13	0 52	43	2 52	13	0.867	43	2.867
14	0 56	44	2 56	14	0.933	44	2.933
15	1 0	45	3 0	15	1.000	45	3.000
16	1 4	46	3 4	16	1.067	46	3.067
17	1 8	47	3 8	17	1.133	47	3.133
18	1 12	48	3 12	18	1.200	48	3.200
19	1 16	49	3 16	19	1.267	49	3.267
20	1 20	50	3 20	20	1.333	50	3.333
21	1 24	51	3 24	21	1.400	51	3.400
22	1 28	52	3 28	22	1.467	52	3.467
23	1 32	53	3 32	23	1.533	53	3.533
24	1 36	54	3 36	24	1.600	54	3.600
25	1 40	55	3 40	25	1.667	55	3.667
26	1 44	56	3 44	26	1.733	56	3.733
27	1 48	57	3 48	27	1.800	57	3.800
28	1 52	58	3 52	28	1.867	58	3.867
29	1 56	59	3 56	29	1.933	59	3.933
30	2 0	60	4 0	30	2.000	60	4.000

To convert Sidereal Time into parts of the Equator in Arc, or to convert Time into Terrestrial Longitude in Arc.

HOURS.		MINUTES.				SECONDS.			
Time	Arc.	Time	Arc.	Time	Arc.	Time	Arc.	Time	Arc.
h	°	m	° ′	m	° ′	s	′ ″	s	′ ″
1	15	1	0 15	31	7 45	1	0 15	31	7 45
2	30	2	0 30	32	8 0	2	0 30	32	8 0
3	45	3	0 45	33	8 15	3	0 45	33	8 15
4	60	4	1 0	34	8 30	4	1 0	34	8 30
5	75	5	1 15	35	8 45	5	1 15	35	8 45
6	90	6	1 30	36	9 0	6	1 30	36	9 0
7	105	7	1 45	37	9 15	7	1 45	37	9 15
8	120	8	2 0	38	9 30	8	2 0	38	9 30
9	135	9	2 15	39	9 45	9	2 15	39	9 45
10	150	10	2 30	40	10 0	10	2 30	40	10 0
11	165	11	2 45	41	10 15	11	2 45	41	10 15
12	180	12	3 0	42	10 30	12	3 0	42	10 30
13	195	13	3 15	43	10 45	13	3 15	43	10 45
14	210	14	3 30	44	11 0	14	3 30	44	11 0
15	225	15	3 45	45	11 15	15	3 45	45	11 15
16	240	16	4 0	46	11 30	16	4 0	46	11 30
17	255	17	4 15	47	11 45	17	4 15	47	11 45
18	270	18	4 30	48	12 0	18	4 30	48	12 0
19	285	19	4 45	49	12 15	19	4 45	49	12 15
20	300	20	5 0	50	12 30	20	5 0	50	13 30
21	315	21	5 15	51	12 45	21	5 15	51	12 45
22	330	22	5 30	52	13 0	22	5 30	52	13 0
23	345	23	5 45	53	13 15	23	5 45	53	13 15
24	360	24	6 0	54	13 30	24	6 0	54	13 30
		25	6 15	55	13 45	25	6 15	55	13 45
		26	6 30	56	14 0	26	6 30	56	14 0
		27	6 45	57	14 15	27	6 45	57	14 15
		28	7 0	58	14 30	28	7 0	58	14 30
		29	7 15	59	14 45	29	7 15	59	14 45
		30	7 30	60	15 0	30	7 30	60	15 0

*To convert Sidereal Time into parts of the Equator in Arc,
or to convert Time into Terrestrial Longitude in Arc.*

TENTHS OF SECONDS.

Time.	Arc.	Time.	Arc.	Time.	Arc.	Time.	Arc.
s	''	s	'	s	''	s	''
0.01	0.15	0.31	4.65	0.61	9.15	0.91	13.65
0.02	0.30	0.32	4.80	0.62	9.30	0.92	13.80
0.03	0.45	0.33	4.95	0.63	9.45	0.93	13.95
0.04	0.60	0.34	5.10	0.64	9.60	0.94	14.10
0.05	0.75	0.35	5.25	0.65	9.75	0.95	14.25
0.06	0.90	0.36	5.40	0.66	9.90	0.96	14.40
0.07	1.05	0.37	5.55	0.67	10.05	0.97	14.55
0.08	1.20	0.38	5.70	0.68	10.20	0.98	14.70
0.09	1.35	0.39	5.85	0.69	10.35	0.99	14.85
0.10	1.50	0.40	6.00	0.70	10.50	1.00	15.00
0.11	1.65	0.41	6.15	0.71	10.65		
0.12	1.80	0.42	6.30	0.72	10.80		
0.13	1.95	0.43	6.45	0.73	10.95		
0.14	2.10	0.44	6.60	0.74	11.10		
0.15	2.25	0.45	6.75	0.75	11.25		Arc.
0.16	2.40	0.46	6.90	0.76	11.40		
0.17	2.55	0.47	7.05	0.77	11.55		
0.18	2.70	0.48	7.20	0.78	11.70		
0.19	2.85	0.49	7.35	0.79	11.85		
0.20	3.00	0.50	7.50	0.80	12.00		''
0.21	3.15	0.51	7.65	0.81	12.15	.001	0.015
0.22	3.30	0.52	7.80	0.82	12.30	.002	0.030
0.23	3.45	0.53	7.95	0.83	12.45	.003	0.045
0.24	3.60	0.54	8.10	0.84	12.60	.004	0.060
0.25	3.75	0.55	8.25	0.85	12.75	.005	0.075
0.26	3.90	0.56	8.40	0.86	12.90	.006	0.090
0.27	4.05	0.57	8.55	0.87	13.05	.007	0.105
0.28	4.20	0.58	8.70	0.88	13.20	.008	0.120
0.29	4.35	0.59	8.85	0.89	13.35	.009	0.135
0.30	4.50	0.60	9.00	0.90	13.50	.010	0.150

The seventh column is headed, vertically, *Thousands of seconds of Time.*

To convert Right Ascension in Arc into Mean Time.

DEGREES.

R. A. in Arc.	Mean Time.	R. A. in Arc.	Mean Time.	R. A. in Arc.	Mean Time.
°	h m s	°	h m s	°	h m s
1	0 3 59.345	31	2 3 39.686	61	4 3 20.027
2	0 7 58.689	32	2 7 39.030	62	4 7 19.371
3	0 11 58.034	33	2 11 38.375	63	5 11 18.716
4	0 15 57.379	34	2 15 37.720	64	4 15 18.061
5	0 19 56.724	35	2 19 37.064	65	4 19 17.405
6	0 23 56.068	36	2 23 36.409	66	4 23 16.750
7	0 27 55.413	37	2 27 35.754	67	4 27 16.095
8	0 31 54.758	38	2 31 35.099	68	4 31 15.639
9	0 35 54.102	39	2 35 34.443	69	4 35 14.784
10	0 39 53.447	40	2 39 33.788	70	4 39 14.129
11	0 43 52.792	41	2 43 33.133	71	4 43 13.474
12	0 47 52.136	42	2 47 32.477	72	4 47 12.818
13	0 51 51.481	43	2 51 31.822	73	4 51 12.163
14	0 55 50.826	44	2 55 31.167	74	4 55 11.508
15	0 59 50.170	45	2 59 30.511	75	4 59 10.852
16	1 3 49.515	46	3 3 29.856	76	5 3 10.197
17	1 7 48.860	47	3 7 29.201	77	5 7 9.542
18	1 11 48.205	48	3 11 28.545	78	5 11 8.886
19	1 15 47.549	49	3 15 27.890	79	5 15 8.231
20	1 19 46.894	50	3 19 27.235	80	5 19 7.576
21	1 23 46.239	51	3 23 26.580	81	5 23 6.920
22	1 27 45.583	52	3 27 25.924	82	5 27 6.265
23	1 31 44.928	53	3 31 25.269	83	5 31 5.610
24	1 35 44.273	54	3 35 24.614	84	5 35 4.955
25	1 39 43.617	55	3 39 23.958	85	5 39 4.299
26	1 43 42.962	56	3 43 23.303	86	5 43 3.644
27	1 47 42.307	57	3 47 22.648	87	5 47 2.989
28	1 51 41.652	58	3 51 21.992	88	5 51 2.333
29	1 55 40.996	59	3 55 21.337	89	5 55 1.678
30	1 59 40.341	60	3 59 20.682	90	5 59 1.023

To convert Right Ascension in Arc into Mean Time.

DEGREES.

R. A. in Arc.	Mean Time.	R. A. in Arc.	Mean Time.	R. A. in Arc.	Mean Time.
°	h m s	°	h m s	°	h m s
91	6 3 0.367	121	8 2 40.708	151	10 2 21.049
92	6 6 59.712	122	8 6 40.053	152	10 6 20.394
93	6 10 59.057	123	8 10 39.398	153	10 10 19.738
94	6 14 58.401	124	8 14 38.742	154	10 14 19.083
95	6 18 57.746	125	8 18 38.087	155	10 18 18.428
96	6 22 57.091	126	8 22 37.432	156	10 22 17.773
97	6 26 56.436	127	8 26 36.776	157	10 26 17.117
98	6 30 55.780	128	8 30 36.121	158	10 30 16.462
99	6 34 55.125	129	8 34 35.466	159	10 34 15.807
100	6 38 54.470	130	8 38 34.810	160	10 38 15.151
101	6 42 53.814	131	8 42 34.155	161	10 42 14.496
102	6 46 53.159	132	8 46 33.500	162	10 46 13.841
103	6 50 52.504	133	8 50 32.845	163	10 50 13.185
104	6 54 51.848	134	8 54 32.189	164	10 54 12.530
105	6 58 51.193	135	8 58 31.534	165	10 58 11.875
106	7 2 50.538	136	9 2 30.879	166	11 2 11.220
107	7 6 49.883	137	9 6 30.223	167	11 6 10.564
108	7 10 49.227	138	9 10 29.568	168	11 10 9.909
109	7 14 48.572	139	9 14 28.913	169	11 14 9.254
110	7 18 47.917	140	9 18 28.257	170	11 18 8.598
111	7 22 47.261	141	9 22 27.602	171	11 22 7.943
112	7 26 46.606	142	9 26 26.947	172	11 26 7.288
113	7 30 45.951	143	9 30 26.292	173	11 30 6.632
114	7 34 45.295	144	9 34 25.636	174	11 34 5.977
115	7 38 44.640	145	9 38 24.981	175	11 38 5.322
116	7 42 43.985	146	9 42 24.326	176	11 42 4.666
117	7 46 43.329	147	9 46 23.670	177	11 46 4.011
118	7 50 42.674	148	9 50 23.015	178	11 50 3.356
119	7 54 42.019	149	9 54 22.360	179	11 54 2.701
120	7 58 41.364	150	9 58 21.704	180	11 58 2.045

To convert Right Ascension in Arc into Mean Time.

MINUTES.				SECONDS.			
R. A. in Arc.	Mean Time.	R. A. in Arc.	Mean Time.	R. A. in Arc.	Mean Time.	R. A. in Arc.	Mean Time.
′	m. s.	′	m. s.	″	s.	″	s.
1	0 3.989	31	2 3.661	1	0.066	31	2.061
2	0 7.978	32	2 7.650	2	0.133	32	2.128
3	0 11.969	33	2 11.640	3	0.199	33	2.194
4	0 15.956	34	2 15.629	4	0.266	34	2.261
5	0 19.945	35	2 19.618	5	0.332	35	2.327
6	0 23.935	36	2 23.607	6	0.399	36	2.393
7	0 27.924	37	2 27.596	7	0.465	37	2.460
8	0 31.913	38	2 31.585	8	0.532	38	2.526
9	0 35.902	39	2 35.574	9	0.598	39	2.593
10	0 39.891	40	2 39.563	10	0.665	40	2.659
11	0 43.880	41	2 43.552	11	0.731	41	2.726
12	0 47.869	42	2 47.541	12	0.798	42	2.792
13	0 51.858	43	2 51.530	13	0.864	43	2.859
14	0 55.847	44	2 55.519	14	0.931	44	2.925
15	0 59.836	45	2 59.509	15	0.997	45	2.992
16	1 3.825	46	3 3.498	16	1.064	46	3.058
17	1 7.814	47	3 7.487	17	1.130	47	3.125
18	1 11.803	48	3 11.476	18	1.197	48	3.191
19	1 15.793	49	3 15.465	19	1.263	49	3.258
20	1 19.782	50	3 19.454	20	1.330	50	3.324
21	1 23.771	51	3 23.443	21	1.396	51	3.391
22	1 27.760	52	3 27.432	22	1.463	52	3.457
23	1 31.749	53	3 31.421	23	1.529	53	3.524
24	1 35.738	54	3 35.410	24	1.596	54	3.590
25	1 93.727	55	3 39.399	25	1.662	55	3.657
26	1 43.716	56	3 43.388	26	1.729	56	3.723
27	1 47.705	57	3 47.377	27	1.795	57	3.790
28	1 51.694	58	3 51.367	28	1.862	58	3.856
29	1 55.683	59	3 55.356	29	1.928	59	3.923
30	1 59.672	60	3 59.345	30	1.995	60	3.989

To convert Mean Time into Right Ascension in Arc.

HOURS.		MINUTES.			
Mean Time.	R. A. in Arc.	Mean Time.	R. A. in Arc.	Mean Time.	R. A. in Arc.
h	° ′ ″	m	° ′ ″	m	° ′ ″
1	15 2 27.85	1	0 15 2.46	31	7 46 16.39
2	30 4 52.69	2	0 30 4.93	32	8 1 18.85
3	45 7 23.54	3	0 45 30.39	33	8 16 21.31
4	60 9 51.39	4	1 0 9.86	34	8 31 23.78
5	75 12 19.24	5	1 15 12.32	35	8 46 26.24
6	90 14 47.08	6	1 30 14.79	36	9 1 28.71
7	105 17 14.93	7	1 45 17.25	37	9 16 31.17
8	120 19 42.78	8	2 0 19.71	38	9 31 33.64
9	135 22 10.62	9	2 15 22.18	39	9 46 36.10
10	150 24 38.47	10	2 30 24.64	40	10 1 38.57
11	165 27 6.32	11	2 45 27.11	41	10 16 41.03
12	180 29 34.16	12	3 0 29.57	42	10 31 43.39
13	195 32 2.01	13	3 15 32.03	43	10 46 45.96
14	210 34 29.86	14	3 30 34.50	44	11 1 48.42
15	225 36 57.70	15	3 45 36.96	45	11 16 50.89
16	240 39 25.55	16	4 0 39.43	46	11 31 53.35
17	255 41 53.40	17	4 15 41.89	47	11 46 55.81
18	270 44 21.24	18	4 30 44.35	48	12 1 58.38
19	285 46 49.09	19	4 45 46.82	49	12 17 0.74
20	300 49 16.94	20	5 0 49.28	50	12 32 3.21
21	315 51 44.78	21	5 15 51.75	51	12 47 5.57
22	330 54 12.63	22	5 30 54.21	52	13 2 8.13
23	345 56 40.48	23	5 45 56.67	53	13 17 10.60
24	360 59 8.33	24	6 0 59.14	54	13 32 13.06
		25	6 16 1.60	55	13 47 15.53
		26	6 31 4.07	56	14 2 17.99
		27	6 46 6.53	57	14 17 20.45
		28	7 1 9.00	58	14 32 22.92
		29	7 16 11.46	59	14 47 25.38
		30	7 31 13.92	60	15 2 27.85

To convert Mean Time into Right Ascension in Arc.

SECONDS AND TENTHS.

Mean Time.	R. A. in Arc.	Mean Time.	R. A in Arc.	Mean Time.	R. A. in Arc.	Mean Time.	R. A. in Arc.
s	′ ″	s	′ ″	s	″	s	″
1	0 15.04	31	7 46.27	0.01	0.15	0.31	4.66
2	0 30.08	32	8 1.31	0.02	0.30	0.32	4.81
3	0 45.12	33	8 16.36	0.03	0.45	0.33	4.96
4	1 0.16	34	8 31.40	0.04	0.60	0.34	5.12
5	1 15.21	35	8 46.44	0.05	0.75	0.35	5.27
6	1 30.25	36	9 1.48	0.06	0.90	0.36	5.42
7	1 45.29	37	9 16.52	0.07	1.05	0.37	5.57
8	2 0.33	38	9 31.56	0.08	1.20	0.38	5.72
9	2 15.37	39	9 46.60	0.09	1.35	0.39	5.87
10	2 30.41	40	10 1.64	0.10	1.50	0.40	6.02
11	2 45.45	41	10 16.68	0.11	1.65	0.41	6.17
12	3 0.49	42	10 31.73	0.12	1.81	0.42	6.32
13	3 15.53	43	10 46.77	0.13	1.96	0.43	6.47
14	3 30.58	44	11 1.81	0.14	2.11	0.44	6.62
15	3 45.62	45	11 16.85	0.15	2.26	0.45	6.77
16	4 0.66	46	11 31.89	0.16	2.41	0.46	6.92
17	4 15.70	47	11 46.93	0.17	2.56	0.47	7.07
18	4 30.74	48	12 1.97	0.18	2.71	0.48	7.22
19	4 45.78	49	12 17.01	0.19	2.86	0.49	7.37
20	5 0.82	50	12 32.05	0.20	3.01	0.50	7.52
21	5 15.86	51	12 47.09	0.21	3.16	0.51	7.67
22	5 30.90	52	13 2.14	0.22	3.31	0.52	7.82
23	5 45.94	53	13 17.18	0.23	3.46	0.53	7.97
24	6 1.00	54	13 32.22	0.24	3.61	0.54	8.12
25	6 16.03	55	13 47.26	0.25	3.76	0.55	8.27
26	6 31.07	56	14 2.30	0.26	3.91	0.56	8.43
27	6 46.11	57	14 17.34	0.27	4.06	0.57	8.58
28	7 1.15	58	14 32.38	0.28	4.21	0.58	8.73
29	7 16.19	59	14 47.42	0.29	4.36	0.59	8.88
30	7 31.23	60	15 2.46	0.30	4.51	0.60	9.03

To convert Mean Time into Right Ascension in Arc.

SECONDS AND TENTHS.

Mean Time.	R. A. in Arc.	Mean Time.	R. A. in Arc.	Mean Time.	R. A. in Arc.	Thousands of Seconds of Mean Time.	R. A. in Arc.
s	//	s	//	s	//		
0.61	9.18	0.76	11.43	0.91	13.69		
0.62	9.33	0.77	11.58	0.92	13.84		
0.63	9.48	0.78	11.74	0.93	13.99		
0.64	9.63	0.79	11.89	0.94	14.14		
0.65	9.78	0.80	12.04	0.95	14.29	s	//
0.66	9.93	0.81	12.19	0.96	14.44	.001	0.02
0.67	10.08	0.82	12.34	0.97	14.59	.002	0.03
0.68	10.23	0.83	12.49	0.98	14.74	.003	0.05
0.69	10.38	0.84	12.64	0.99	14.89	.004	0.06
0.70	10.53	0.85	12.79	1.00	15.05	.005	0.08
0.71	10.68	0.86	12.94			.006	0.09
0.72	10.83	0.87	13.09			.007	0.11
0.73	10.98	0.88	13.24			.008	0.12
0.74	11.13	0.89	13.39			.009	0.14
0.75	11.28	0.90	13.54			.010	0.15

	Logarithms.
12 hours, expressed in seconds = 43200.	4.6354837
Complement to the same = .00002315	5.3645163
24 hours, expressed in seconds = 86400.	4.9365137
Complement to the same = .00001157	5.0634863
360 degrees, expressed in seconds = 1296000	6.1126050
To convert Sidereal time to M. solar time −	9.9988126

FORM FOR

SURVEY OF   DETERMINATION OF TIME,

DATE AND STATION.—1843, *October* 13—*Mouth of the Big Black river,*

INSTRUMENTS. { Sextant No. 2197, by *Troughton & Simms*, and *Mean Solar* Chronometer No. 76, by *Charles*

NAMES OF STARS.	Observed double altitudes of Star with Sextant.	True altitudes of Star affected by corrections for refraction and errors of Sextant = A.	Mean Solar time of observation deduced.	Time of observation noted by Chronometer.
	° ′ ″	° ′ ″	h. m. s.	h. m. s.
	91 43 40	45 52 58.8	7 05 47.69	6 57 02.4
	92 18 00	46 10 09.3	7 07 28.67	6 58 43.2
α Andromedæ	92 41 15	46 21 47.3	7 08 37.15	6 59 52.8
	93 04 05	46 33 12.6	7 09 44.37	7 00 59.6
(*East.*)	93 45 20	46 53 50.8	7 11 45.92	7 03 01.2
	94 13 45	47 08 03.7	7 13 09.73	7 04 25.6
	94 40 50	47 21 36.6	7 14 29.64	7 05 45
	95 07 25	47 34 54.5	7 15 48.14	7 07 03.6

Mean result of 8 observations on *α Andromedæ*, in the *East*,

NAMES OF STARS.	Observed double altitudes of Star with Sextant.	True altitudes of Star affected by corrections for refraction and errors of Sextant = A.	Mean Solar time of observation deduced.	Time of observation noted by Chronometer.
	° ′ ″	° ′ ″	h. m. s.	h. m. s.
	95 20 05	47 41 14.7	8 55 32.36	8 46 49.2
	95 00 00	47 31 11.6	8 56 32.06	8 47 50.4
α Lyræ	94 30 40	47 16 31.2	8 57 59.42	8 49 16
	94 12 20	47 07 21	8 58 54	8 50 10.8
(*West.*)	93 53 45	46 58 03.1	8 59 49.4	8 51 06.9
	93 29 20	46 45 50.2	9 01 02.1	8 52 19.4
	93 07 35	46 34 57.8	9 02 07	8 53 24.8
	92 46 50	46 24 34.5	9 03 09	8 54 26
	92 28 45	46 15 31.7	9 04 02.96	8 55 21.2

Mean result of *nine* observations on the Star *α Lyræ* in the *West* . . . .

Mean result of *eight* observations on the Star *α Andromedæ* in the East as above .

CHRONOMETER ERROR.—*Slow* of *Mean Solar* time *at* 8 *h. p. m.,* by a mean of these results from E. and W. Stars . . . .

RECORD AND COMPUTATION.

by observed double altitudes of East and West Stars.

a tributary to the river St. John, Maine.
artificial horizon of Mercury.
Young.

Chronometer (*C. Y.* 76) slow of mean solar time by each observation.	REMARKS.
h. m. s. 0 08 45.29 8 45.47 8 44.35 8 44.77 8 44.72 8 44.13 8 44.64 8 44.54	Index error of Sextant $\quad=\quad +2' 40''$ Error of eccentricity of Sextant $\quad=\quad +1' 32''$ Thermometer 31°.5 Fahr. Barometer 29.14 inches. Apparent AR. of Star, $\quad= 0^h 00^m 21^s.72$ Apparent declination of Star, $= 28° 13'\ 59''.5$ N. Appt. N. Polar distance of Star $= 61\ 46\ \ 00\ .5 = \triangle$ Approximate lat. of this Station $= 46\ 57\ \ 00$ N. $= $ L Approximate longitude of do. $\quad= 4^h 37^m 47^s$ Sider'l time of mean noon at Station $= 13\ 26\ \ 20\ .83$
$0^h 08^m 44^s.74$	
h. m. s. 0 08 43.16 8 41.66 8 43.42 8 43.20 8 42.50 8 42.70 8 42.20 8 43.00 8 41.76	Thermometer 29° Fahr. Barometer 29.14 inches. Apparent A. R. of Star, $\quad= 18^h 31^m 39^s.16$ Apparent declination of Star N. $= 38° 38'\ 46''.5$ App't N. Polar distance of Star $= 51\ 21\ \ 13\ .5\ = \triangle$ Index error of Sextant $\quad=\quad +2' 40''$ Error of eccentricity of Sextant $\quad=\quad +1' 32''$
$0^h 08^m 42^s.6$ 0 08 44.7	
$0^h 08^m 43^s.6$	Observer, *Major J. D. Graham.* Computer, *Private F. Herbst.*

Computation of the 5th of the preceding altitudes of α An-
dromeda. Formula page 143.

Observed double altitude	$= 93° \, 45' \, 20''$
Index error, sextant	$= \quad + \, 2 \; 40$
Excentricity, sextant	$= \quad + \, 1 \; 32$
Double altitude, corrected	$= 93 \; 49 \; 32$
Altitude	$= 46 \; 54 \; 46$
Refraction (Therm. 31°.5—Bar. 29.1)	$= \quad\quad -\, 55.2$
True altitude of $\ast$ $\quad = A$	$= 46° \, 53' \, 50''.8$

$$2m = L + \triangle + A$$

$L =$	$46° \, 57'$	$\ldots\ldots$ Cos $\ldots$	$= 9.8341894$
$\triangle =$	$61 \; 46 \; 00.5$	$\ldots\ldots$ Sin $\ldots$	$= 9.9449899$
$A =$	$46 \; 53 \; 50.8$	Cos L . Sin $\triangle$	$= 9.7791793$

$2m =$	$155° \, 36' \, 51''.3$		
$m =$	$77 \; 48 \; 25 \; .6$	$\ldots\ldots$ Cos $\ldots\ldots$	$= 9.3247069$
$(m - A) =$	$30 \; 54 \; 34 \; .8$	$\ldots\ldots$ Sin $\ldots\ldots$	$= 9.7106984$
		Cos m Sin $(m - A)$	$= 9.0354053$

$$\text{Sin}^2 \tfrac{1}{2} p = \frac{\text{Cos } m \, . \, \text{Sin } (m - A)}{\text{Cos } L \, . \, \text{Sin } \triangle} \quad \ldots\ldots \; . \quad = 19.2562259$$

	Sin $\frac{1}{2} p$	$= \quad 9.6281129$
	$\frac{1}{2} p$	$= 25° \, 08' \, 00''.7$
	p in arc	$= 50 \; 16 \; 01.4$
(page 146)	p in time	$= - \; 3^h \; 21^m \; 04^s.09$
	AR . $\ast$	$= \quad 24 \; 00 \; \; 21.72$
Sidereal time of observation	$= AR \pm p$ $\quad =$	$20 \; 39 \; \; 17.63$
Sidereal time, mean noon, at place, (Naut. Alm.) $=$		$13 \; 26 \; \; 20.83$
Sidereal interval past mean noon	$=$	$7 \; 12 \; \; 56.8$
Retardation of mean on sidereal interval, (page 144) $=$		$- \; 1 \; \; 10.9$

Mean solar interval past mean noon, or mean time, P. M., of observation..............	$7^h \; 11^m \; 45^s.9$
Time of observation by Chronometer	$7 \; 03 \; \; 01.2$
Chronometer slow	$8^m \; 44^s.7$

III. *To find the time by equal Altitudes of the Sun.*

Correction in time, to be applied as an equation to the mean of the times of observed equal altitudes of the sun, in order to obtain the time of its meridional passage.

$$x = \frac{\delta}{48^{\mathrm{h}}} \times \frac{\mathrm{T}}{30} \left(\frac{\mathrm{tang\ D}}{\mathrm{tang\ } 7\frac{1}{2}\ \mathrm{T}} - \frac{\mathrm{tang\ L}}{\mathrm{sin\ } 7\frac{1}{2}\ \mathrm{T}} \right)$$

$$= \delta . \mathrm{tang\ D} \frac{\mathrm{T}}{1440\ \mathrm{tang\ } 7\frac{1}{2}\ \mathrm{T}} - \delta\ \mathrm{tang\ L} \frac{\mathrm{T}}{1440\ \mathrm{sin\ } 7\frac{1}{2}\ \mathrm{T}}$$

make

$$\frac{\mathrm{T}}{1440\ \mathrm{sin\ } 7\frac{1}{2}\ \mathrm{T}} = \mathrm{A}$$

$$\frac{\mathrm{T}}{1440\ \mathrm{tang\ } 7\frac{1}{2}\ \mathrm{T}} = \mathrm{B}$$

$$x = \mp \mathrm{A} . \delta . \mathrm{tang\ L} + \mathrm{B} . \delta . \mathrm{tang\ D}$$

$$\text{apparent noon} = \tfrac{1}{2}\ (t + t') + x$$

$t, t' = $ the times of observation.

 $\mathrm{T} = (t' - t) = $ the interval of time between the obser- vations, expressed in hours and decimals.

 $\mathrm{L} = $ the latitude of the place of observation : (*minus* when south.)

 $\mathrm{D} = $ the declination, at the time of noon, on the given day: (*minus* when south.)

 $\delta = $ the double daily variation in the declination, de- duced from the noon of the preceding day to noon of the following day: (*minus* when the Sun is proceeding towards the south.)

 $x = $ required correction in *seconds*: where A is to be *minus* when time of noon is required, and *plus* when time of midnight is required, *i. e.* when the first observation is made in the afternoon, and the corresponding one the morning following.

Log. values of A and B are given in the tables.

21

Equations to equal Altitudes.

Interval.	Log A.	Log B.	Interval.	Log A.	Log B.
H. M.			H. M.		
2 0	7.7297	7.7146	3 0	7.7359	7.7015
2	.7298	.7143	2	.7362	.7010
4	.7300	.7139	4	.7364	.7005
6	.7302	.7136	6	.7367	.6999
8	.7304	.7132	8	.7369	.6993
10	.7305	.7128	10	.7372	.6988
12	.7307	.7125	12	.7374	.6982
14	.7309	.7121	14	.7377	.6976
16	.7311	.7117	16	.7380	.6970
18	.7313	.7113	18	.7383	.6964
20	.7315	.7109	20	.7386	.6958
22	.7317	.7105	22	.7388	.6952
24	.7319	.7101	24	.7391	.6946
26	.7321	.7097	26	.7394	.6940
28	.7323	.7092	28	.7397	.6934
30	.7325	.7088	30	.7400	.6927
32	.7327	.7083	32	.7403	.6921
34	.7329	.7079	34	.7406	.6914
36	.7331	.7075	36	.7409	.6908
38	.7333	.7070	38	.7412	.6901
40	.7336	.7065	40	.7415	.6894
42	.7338	.7061	42	.7418	.6888
44	.7340	.7056	44	.7421	.6881
46	.7342	.7051	46	.7424	.6874
48	.7345	.7046	48	.7428	.6867
50	.7347	.7041	50	.7431	.6859
52	.7349	.7036	52	.7434	.6852
54	.7352	.7031	54	.7437	.6845
56	.7354	.7026	56	.7441	.6838
58	7.7357	7.7021	58	7.7444	7.6830

$$x = \mp A\,\delta\,\text{tang}\,L + B\,\delta\,\text{tang}\,D.$$

Equations to equal Altitudes.

Interval.	Log A.	Log B.	Interval.	Log A.	Log B.
H. M.			H. M.		
4 0	7.7447	7.6823	5 0	7.7562	7.6556
2	.7451	.6815	2	.7566	.6546
4	.7454	.6807	4	.7570	.6536
6	.7458	.6800	6	.7575	.6525
8	.7461	.6792	8	.7579	.6514
10	.7464	.6784	10	.7583	.6504
12	.7468	.6776	12	.7588	.6493
14	.7472	.6768	14	.7592	.6482
16	.7475	.6759	16	.7597	.6471
18	.7479	.6751	18	.7601	.6460
20	.7482	.6743	20	.7606	.6448
22	.7486	.6734	22	.7610	.6437
24	.7490	.6726	24	.7615	.6425
26	.7494	.6717	26	.7620	.6414
28	.7497	.6708	28	.7624	.6402
30	.7501	.6700	30	.7629	.6390
32	.7505	.6691	32	.7634	.6378
34	.7509	.6682	34	.7638	.6366
36	.7513	.6673	36	.7643	.6354
38	.7517	.6663	38	.7648	.6342
40	.7521	.6654	40	.7653	.6329
42	.7525	.6645	42	.7658	.6317
44	.7529	.6635	44	.7663	.6304
46	.7533	.6626	46	.7668	.6291
48	.7537	.6616	48	.7673	.6278
50	.7541	.6606	50	.7678	.6265
52	.7545	.6597	52	.7683	.6252
54	.7549	.6587	54	.7688	.6239
56	.7553	.6577	56	.7693	.6225
58	7.7557	7.6567	58	7.7698	7.6212

$$x = \mp \, A \, \delta \, \text{tang } L + B \, \delta \, \text{tang } D.$$

Equations to Equal Altitudes.

Interval.	Log A.	Log B.	Interval.	Log A.	Log B.
H. M.			H. M.		
6 0	7.7703	7.6198	7 0	7.7873	7.5717
2	.7708	.6184	2	.7879	.5699
4	.7713	.6170	4	.7885	.5680
6	.7719	.6156	6	.7891	.5661
8	.7724	.6142	8	.7898	.5641
10	.7729	.6127	10	.7904	.5622
12	.7735	.6113	12	.7910	.5602
14	.7740	.6098	14	.7916	.5582
16	.7745	.6083	16	.7923	.5562
18	.7751	.6068	18	.7929	.5542
20	.7756	.6053	20	.7936	.5522
22	.7762	.6038	22	.7942	.5501
24	.7767	.6023	24	.7949	.5480
26	.7773	.6007	26	.7955	.5459
28	.7779	.5991	28	.7962	.5437
30	.7784	.5975	30	.7969	.5416
32	.7790	.5959	32	.7975	.5394
34	.7796	.5943	34	.7982	.5372
36	.7801	.5927	36	.7989	.5350
38	.7807	.5910	38	.7995	.5327
40	.7813	.5894	40	.8002	.5304
42	.7819	.5877	42	.8009	.5281
44	.7825	.5860	44	.8016	.5258
46	.7831	.5843	46	.8023	.5234
48	.7836	.5825	48	.8030	.5211
50	.7842	.5808	50	.8037	.5186
52	.7848	.5790	52	.8044	.5162
54	.7854	.5772	54	.8051	.5137
56	.7860	.5754	56	.8058	.5112
58	7.7867	7.5736	58	7.8065	7.5087

$$x = \mp\, A\, \delta\, \tan L + B\, \delta\, \tan D.$$

Equations to Equal Altitudes.

Interval.	Log A.	Log B.	Interval.	Log A.	Log B.
H. M.			H. M.		
8 0	7.8072	7.5062	15 0	8.0521	—7.6350
2	.8079	.5036	2	.0539	.6413
4	.8086	.5010	4	.0556	.6475
6	.8094	.4983	6	.0574	.6537
8	.8101	.4957	8	.0592	.6599
10	.8108	.4930	10	.0610	.6660
12	.8116	.4902	12	.0628	.6721
14	.8123	.4874	14	.0646	.6781
16	.8130	.4846	16	.0664	.6841
18	.8138	.4818	18	.0682	·6900
20	.8145	.4789	20	.0700	.6959
22	.8153	.4760	22	.0718	.7018
24	.8160	.4731	24	.0737	.7077
26	.8168	.4701	26	.0755	.7135
28	.8176	.4671	28	.0774	.7192
30	.8183	.4640	30	.0792	.7249
32	.8191	.4609	32	.0811	.7306
34	.8199	.4578	34	.0830	.7363
36	.8206	.4546	36	.0849	.7419
38	.8214	.4514	38	.0868	.7475
40	.8222	.4482	40	.0887	.7531
42	.8230	.4449	42	.0906	.7586
44	.8238	.4415	44	.0925	.7641
46	.8246	.4381	46	.0945	.7696
48	.8254	.4347	48	.0964	.7751
50	.8262	.4312	50	.0983	.7805
52	.8270	.4277	52	.1003	.7859
54	.8278	.4241	54	.1023	.7912
56	.8286	.4205	56	.1042	.7966
58	7.8294	7.4168	58	8.1062	—7.8019

$$x = \mp A\, \delta\, \text{tang}\, L + B\, \delta\, \text{tang}\, D.$$

Equations to Equal Altitudes.

Interval.	Log A.	Log B.	Interval.	Log A.	Log B.
H. M.			H. M.		
16 0	8.1082	—7.8072	17 0	8.1726	—7.9571
2	.1102	.8125	2	.1749	.9618
4	.1122	.8177	4	.1773	.9666
6	.1143	.8229	6	.1796	.9713
8	.1163	.8281	8	.1819	.9761
10	.1183	.8333	10	.1843	.9808
12	.1204	.8385	12	.1867	.9855
14	.1224	.8436	14	.1890	.9902
16	.1245	.8487	16	.1914	.9949
18	.1266	.8538	18	.1938	—7.9996
20	.1287	.8589	20	.1963	—8.0043
22	.1308	.8640	22	.1987	.0090
24	.1329	.8690	24	.2011	.0137
26	.1350	.8740	26	.2036	.0184
28	.1371	.8790	28	.2061	.0230
30	.1393	.8840	30	.2086	.0277
32	.1414	.8890	32	.2111	.0323
34	.1436	.8939	34	.2136	.0370
36	.1458	.8989	36	.2161	.0416
38	.1479	.9038	38	.2186	.0462
40	.1501	.9087	40	.2212	.0508
42	.1523	.9136	42	.2237	.0555
44	.1545	.9185	44	.2263	.0601
46	.1568	.9234	46	.2289	.0647
48	.1590	.9282	48	.2315	.0693
50	.1612	.9330	50	.2341	.0739
52	.1635	.9379	52	.2367	.0785
54	.1658	.9427	54	.2394	.0831
56	.1680	.9475	56	.2420	.0877
58	8.1703	—7.9523	58	8.2447	—8.0923

$$x = \mp\, A\, \delta\, \tan L + B\, \delta\, \tan D.$$

Equations to Equal Altitudes.

Interval.	Log A.	Log B.	Interval.	Log A.	Log B.
H. M.			H. M.		
18 0	8.2474	—8.0969	19 0	8.3359	—8.2354
2	.2501	.1015	2	.3392	.2401
4	.2529	.1061	4	.3424	.2448
6	.2556	.1107	6	.3457	.2495
8	.2583	.1153	8	.3490	.2542
10	.2611	.1199	10	.3524	.2589
12	.2639	.1245	12	.3557	.2637
14	.2667	.1291	14	.3591	.2684
16	.2695	.1336	16	.3625	.2732
18	.2723	.1382	18	.3659	.2779
20	.2752	.1428	20	.3694	.2827
22	.2781	.1474	22	.3728	.2875
24	.2809	.1520	24	.3763	.2923
26	.2838	.1566	26	.3798	.2971
28	.2868	.1612	28	.3834	.3019
30	.2897	.1658	30	.3869	.3068
32	.2926	.1704	32	.3905	.3116
34	.2956	.1750	34	.3941	.3165
36	.2986	.1797	36	.3978	.3214
38	.3016	.1842	38	.4015	.3263
40	.3046	.1889	40	.4052	.3312
42	.3077	.1935	42	.4089	.3361
44	.3107	.1981	44	.4126	.3410
46	.3138	.2028	46	.4164	.3460
48	.3169	.2074	48	.4202	.3510
50	.3200	.2121	50	.4241	.3560
52	.3232	.2167	52	.4279	.3610
54	.3263	.2214	54	.4318	.3660
56	.3295	.2261	56	.4357	.3711
58	8.3327	—8.2307	58	8.4397	—8.3761

$$x = \mp \text{ A } \delta \text{ tang L} + \text{ B } \delta \text{ tang D.}$$

Equations to Equal Altitudes.

Interval.	Log A.	Log B.	Interval.	Log A.	Log B.
H. M.			H. M.		
20 0	8.4437	—8.3812	21 0	8.5810	—8.5466
2	.4477	.3863	2	.5863	.5527
4	.4518	.3915	4	.5917	.5588
6	.4559	.3966	6	.5971	.5650
8	.4600	.4018	8	.6025	.5712
10	.4641	.4070	10	.6081	.5775
12	.4683	.4122	12	.6136	.5838
14	.4726	.4175	14	.6193	.5902
16	.4768	.4227	16	.6250	.5966
18	.4811	.4280	18	.6308	.6031
20	.4854	.4334	20	.6366	.6096
22	.4898	.4387	22	.6426	.6162
24	.4942	.4441	24	.6486	.6229
26	.4987	.4495	26	.6546	.6296
28	.5032	.4549	28	.6608	.6364
30	.5077	.4604	30	.6670	.6433
32	.5123	.4659	32	.6733	.6502
34	.5169	.4714	34	.6796	.6572
36	.5215	.4770	36	.6861	.6643
38	.5262	.4826	38	.6927	.6715
40	.5310	.4882	40	.6993	.6788
42	.5357	.4939	42	.7060	.6860
44	.5406	.4996	44	.7128	.6934
46	.5455	.5053	46	.7197	.7009
48	.5504	.5111	48	.7268	.7085
50	.5554	.5169	50	.7339	.7162
52	.5604	.5228	52	.7411	.7239
54	.5655	.5287	54	.7484	.7318
56	.5706	.5346	56	.7558	.7398
58	8.5758	—8.5406	58	8.7634	—8.7478

$$x = \mp \, A \, \delta \, \tan g \, L + B \, \delta \, \tan g \, D.$$

Equations to Equal Altiutdes.

Interval.	Log A.	Log B.	Interval.	Log A.	Log B.
H. M.			H. M.		
22 0	8.7711	—8.7560	23 0	9.0877	—9.0839
2	.7789	.7643	2	.1029	.0995
4	.7868	.7727	4	.1187	.1155
6	.7948	.7813	6	.1351	.1321
8	.8030	.7899	8	.1520	.1492
10	.8113	.7987	10	.1696	.1670
12	.8198	.8076	12	.1879	.1855
14	.8284	.8167	14	.2069	.2047
16	.8372	.8259	16	.2268	.2248
18	.8461	.8353	18	.2476	.2456
20	.8553	.8448	20	.2693	.2677
22	.8645	.8545	22	.2922	.2907
24	.8740	.8644	24	.3162	.3149
26	.8837	.8745	26	.3416	.3404
28	.8935	.8847	28	.3685	.3674
30	.9036	.8952	30	.3971	.3962
32	.9139	.9058	32	.4276	.4268
34	.9244	.9167	34	.4604	.4597
36	.9351	.9278	36	.4957	.4952
38	.9461	.9391	38	.5341	.5336
40	.9574	.9507	40	.5761	.5757
42	.9689	.9626	42	.6224	.6221
44	.9807	.9747	44	.6742	.6739
46	8.9928	.9871	46	.7328	.7326
48	9.0052	—8.9999	48	.8003	.8001
50	.0180	—9.0129	50	.8801	.8800
52	.0311	.0263	52	9.9776	—9.9775
54	.0446	.0401	54	0.1031	—0.1031
56	.0585	.0543	56	.2798	.2798
58	9.0729	—9.0689	58	0.5814	—0.5814

$$x = \mp \text{ A } \delta \text{ tang L} + \text{B } \delta \text{ tang D.}$$

22

SURVEY OF   DETERMINATION OF THE TIME,
Chronometer

DATE AND STATION.—1844, *August 9—American Camp, near Tasche-*

INSTRUMENTS. { Sextant No. 2197, by *Troughton & Simms*, and *Mean Solar* Chronometer No. 2440, by *Parkinson*

Observed double altitudes of the Sun's upper and lower limbs.	Times by Chronometer, of observed equal altitudes. *August 9th.*		$t' - t =$ the elapsed time, $= T$.	Equation of equal altitudes $= x$.
	A. M. $= t$.	P. M. $= t'$.		
Upper Limb.	h. m. s.	h. m. s.	h. m.	s.
78° 50′ 00″	1 28 23	8 03 16.5		
79 19 30	1 28 52.8	8 01 46.5	6 33	+10.63
Lower Limb.				
83° 10′ 00″	1 45 01	7 46 40.5		
83 40 00	1 46 34.5	7 45 06.2	5 59½	+10.24
84 00 00	1 47 38	7 44 04		
Upper Limb.				
85° 36′ 00″	1 49 23	7 42 18		
87 02 10	1 53 55.5	7 37 46.2	5 48	+10.1

CHRONOMETER ERROR.—*Fast* of mean solar time at apparent noon of *August* 9, 1844, by a mean of 7 pairs of equal altitudes of the Sun,

by observed equal altitudes of the Sun's limbs, to correct the
at noon.

reau's house, on the highland boundary between Maine and Canada.
artificial horizon of Mercury.
& Frodsham.

Chronometer *No.* 2440, *Fast* of mean time at appt. noon, by each pair of equal altitudes.	REMARKS.
h. m. s.	Index error of Sextant . . . . .
4 40 51.29	Error of excentricity of Sextant . . .
4 40 51.17	Thermr. (A. M.) 70° Fahr. Barom. . .
	Thermr. (P. M.) 69° Fahr. Barom. . .
4 40 51.9	Sun's appt. declination at appt. noon (D) $= 15°43'\,12''$ N.
4 40 51.5	Double daily variation of Sun's declination (δ) $= 34'\,54''$
4 40 52.15	$= 2094''$ in arc.
	Equation of time at apparent noon $+ 5^{\mathrm{m}}\,09^{\mathrm{s}}.09$
4 40 51.51	Latitude of station (approximate) $+ 45°\,48' = $ (L.)
4 40 51.86	
$4^{\mathrm{h}}\,40^{\mathrm{m}}\,51^{\mathrm{s}}.6$	Observer, *Major J. D. Graham,* Computer, *Do.*

Computation of the equation of equal altitudes to correct the chronometer for noon of August 9, 1844, by the first of the preceding equal altitudes of the Sun's upper and lower limbs.

$$x = (-\text{A}.\,\delta.\,\text{tang L}) + (\text{B}.\,\delta.\,\text{tang D}.)$$

1st Set.

T = 6ʰ 33ᵐ, log A (page 164) = — 7.77930	log B	= 7.59510
δ = 2094, log δ = — 3.32097	log δ	= — 3.32097
L = 45° 48′, log tang = + 0.01213	log tang D =	9.44933

1st term = + 12ˢ.95   = + 1.11240   — 2ˢ.32   = — 0.36540
2d term = — 2.32

x = + 10ˢ.63 Equation of equal altitudes.

Computation of the first two of the foregoing pairs of equal altitudes of the Sun's limbs.

			1st pair.	2d pair.
A. M.	=	t =	1ʰ 28ᵐ 23ˢ.0	1ʰ 29ᵐ 52ˢ.8
P. M.	=	t' =	8 03 16 .5	8 01 46 .5
		$t + t'$ =	9 31 39 .5	9 31 39 .3
		$\dfrac{t + t'}{2}$ =	4 45 49 .75	4 45 39 .65
Equat'n of equal altitudes =		x =	+ 10 .63	10 .63
Time by chron. of appt. noon		=	4 46 00 .38	4 46 00 .28
Correct mean time at apparent noon (Naut. Alm.)		=	0 05 09 .09	0 05 09 .09
Chron. fast of mean time at appt. noon, August 9, 1844		=	4ʰ 40ᵐ 51ˢ.29	4ʰ 40ᵐ 51ˢ.17

Sun's Parallax in Altitude.

Sun's Altit.	Sun's Horizontal Parallax.					Sun's Altit.	Sun's Horizontal Parallax.				
	$8''.4$	$8''.5$	$8''.6$	$8''.7$	$8''.8$		$8''.4$	$8''.5$	$8''.6$	$8''.7$	$8''.8$
°						°					
0	8.40	8.50	8.60	8.70	8.80	45	5.94	6.01	6.08	6.15	6.22
5	8.37	8.47	8.57	8.67	8.77	50	5.40	5.46	5.53	5.59	5.66
10	8.27	8.37	8.47	8.57	8.67	55	4.82	4.88	4.93	4.99	5.05
15	8.11	8.21	8.31	8.40	8.50	60	4.20	4.25	4.30	4.35	4.40
20	7.89	7.99	8.08	8.18	8.27	65	3.55	3.59	3.63	3.68	3.72
25	7.61	7.70	7.79	7.88	7.98	70	2.87	2.91	2.94	2.98	3.01
30	7.28	7.36	7.45	7.53	7.62	75	2.17	2.20	2.23	2.25	2.28
35	6.88	6.96	7.04	7.13	7.21	80	1.46	1.48	1.49	1.51	1.53
40	6.44	6.51	6.59	6.66	6.74	85	0.73	0.74	0.75	0.76	0.77
45	5.94	6.01	6.08	6.15	6.22	90	0.00	0.00	0.00	0.00	0.00

Parallax in Altitude = Hor. Par $\times$ Cos. Altitude.

Decimals of an Hour.

	MINUTES.						SECONDS.				
M.	Decm.	M.	Decm.	M.	Decm.	s.	Decm.	s.	Decm.	s.	Decm.
1	.01667	21	.35000	41	.68333	1	.00028	21	.00583	41	.01139
2	.03333	22	.36667	42	.70000	2	.00056	22	.00611	42	.01167
3	.05000	23	.38333	43	.71667	3	.00083	23	.00639	43	.01194
4	.06667	24	.40000	44	.73333	4	.00111	24	.00667	44	.01222
5	.08333	25	.41667	45	.75000	5	.06139	25	.00694	45	.01250
6	.10000	26	.43333	46	.76667	6	.00167	26	.00722	46	.01278
7	.11667	27	.45000	47	.78333	7	.00194	27	.00750	47	.01306
8	.13333	28	.46667	48	.80000	8	.00222	28	.00778	48	.01333
9	.15000	29	.48333	49	.81667	9	.00250	29	.00806	49	.01361
10	.16667	30	.50000	50	.83333	10	.00278	30	.00833	50	.01389
11	.18333	31	.51667	51	.85000	11	.00306	31	.00861	51	.01417
12	.20000	32	.53333	52	.86667	12	.00333	32	.00889	52	.01444
13	.21667	33	.55000	53	.88333	13	.00361	33	.00917	53	.01472
14	.23333	34	.56667	54	.90000	14	.00389	34	.00944	54	.01500
15	.25000	35	.58333	55	.91667	15	.00417	35	.00972	55	.01528
16	.26667	36	.60000	56	.93333	16	.00444	36	.01000	56	.01556
17	.28333	37	.61667	57	.95000	17	.00472	37	.01028	57	.01583
18	.30000	38	.63333	58	.96667	18	.00500	38	.01056	58	.01611
19	.31667	39	.65000	59	.98333	19	.00528	39	.01083	59	.01639
20	.33333	40	.66667	60	1.00000	20	.00556	40	.01111	60	.01667

Fahrenheit's Thermometer 50°.　　Barometer 30 Inches.

Alt.	r.	Log. r.	Diff.	Alt.	r.	Log. r.	Diff.
° ′	′ ″			° ′	′ ″		
90 00	0 0.00	0.0000		83 00	0 7.17	0.8557	102
89 50	0.17	9.2304	3011	82 50	7.34	0.8659	101
40	0.34	9.5315	1761	40	7.52	0.8760	99
30	0.51	9.7076	1249	30	7.69	0.8859	97
20	0.68	9.8325	969	20	7.86	0.8956	95
10	0.85	9.9294	791	10	8.04	0.9051	93
89 00	0 1.02	0.0085	670	82 00	0 8.21	0.9144	90
88 50	1.19	0.0755	580	81 50	8.38	0.9234	89
40	1.36	0.1335	512	40	8.56	0.9323	87
30	1.53	0.1847	457	30	8.73	0.9410	85
20	1.70	0.2304	414	20	8.90	0.9495	84
10	1.87	0.2718	379	10	9.08	0.9579	84
88 00	0 2.04	0.3097	347	81 00	0 9.25	0.9663	80
87 50	2.21	0.3444	322	80 50	9.42	0.9743	80
40	2.38	0.3766	301	40	9.60	0.9823	78
30	2.55	0.4067	280	30	9.77	0.9901	77
20	2.72	0.4347	263	20	9.95	0.9978	76
10	2.89	0.4610	250	10	10.12	1.0054	75
87 00	0 3.06	0.4860	235	80 00	10.30	1.0129	72
86 50	3.23	0.5095	224	79 50	0 10.47	1.0201	72
40	3.40	0.5319	211	40	10.65	1.0273	71
30	3.57	0.5530	203	30	10.82	1.0344	70
20	3.74	0.5733	193	20	11.00	1.0414	69
10	3.91	0.5926	186	10	11.17	1.0483	69
86 00	0 4.08	0.6112	178	79 00	0 11.35	1.0552	66
85 50	4.26	0.6290	171	78 50	11.53	1.0618	66
40	4.43	0.6461	165	40	11.71	1.0684	66
30	4.60	0.6626	158	30	11.89	1.0750	65
20	4.77	0.6784	153	20	12.06	1.0815	64
10	4.94	0.6937	149	10	12.24	1.0879	62
85 00	0 5.11	0.7086	142	78 00	0 12.42	1.0941	62
84 50	5.28	0.7228	139	77 50	12.60	1.1003	61
40	5.45	0.7367	135	40	12.78	1.1064	60
30	5.63	0.7502	131	30	12 95	1.1124	60
20	5.80	0.7633	127	20	13.13	1.1184	58
10	5.97	0.7760	122	10	13.31	1.1242	58
84 00	0 6.14	0.7882	120	77 00	0 13.49	1.1300	57
83 50	6.31	0.8002	116	76 50	13.67	1.1357	57
40	6.48	0.8118	114	40	13.85	1.1414	55
30	6.66	0.8232	111	30	14.02	1.1469	55
20	6.83	0.8343	108	20	14.20	1.1524	54
10	7.00	0.8451	106	10	14.38	1.1578	54
83.00	0 7.17	0.8557	102	76.00	0 14.56	1.1632	54

Fahrenheit's Thermometer 50°.   Barometer 30 Inches.

Alt.	r.	Log. r.	Diff.	Alt.	r.	Log. r.	Diff.
° ′	′ ″			° ′	′ ″		
76 00	0 14.56	1.1632		69 00	0 22.42	1.3507	
			54				37
75 50	14.74	1.1686		68 50	22.62	1.3544	
			54				38
40	14.93	1.1740		40	22.81	1.3582	
			53				37
30	15.11	1.1793		30	23.01	1.3619	
			52				37
20	15.29	1.1845		20	23.21	1.3656	
			52				37
10	15.48	1.1897		10	23.40	1.3693	
			50				36
75 00	0 15.66	1.1947		68 00	0 23.60	1.3729	
			51				37
74 50	15.84	1.1998		67 50	23.80	1.3766	
			50				36
40	16.03	1.2048		40	24.00	1.3802	
			50				36
30	16.21	1.2098		30	24.20	1.3838	
			49				36
20	16.39	1.2147		20	24.40	1.3874	
			48				35
10	16.58	1.2195		10	24.60	1.3909	
			46				36
74 00	0 16.75	1.2241		67 00	0 24.80	1.3945	
			46				36
73 50	16.93	1.2287		66 50	25.00	1.3981	
			47				34
40	17.12	1.2334		40	25.20	1.4015	
			46				34
30	17.30	1.2380		30	25.41	1.4049	
			46				35
20	17.48	1.2426		20	25.61	1.4084	
			46				34
10	17.67	1.2472		10	25.81	1.4118	
			47				33
73 00	0 17.86	1.2519		66 00	0 26.01	1.4151	
			45				34
72 50	18.05	1.2564		65 50	26.21	1.4185	
			45				34
40	18.23	1.2609		40	26.42	1.4219	
			44				34
30	18.42	1.2653		30	26.62	1.4253	
			44				33
20	18.61	1.2697		20	26.83	1.4286	
			43				33
10	18.79	1.2740		10	27.03	1.4319	
			44				33
72 00	0 18.98	1.2784		65 00	0 27.24	1.4352	
			42				33
71 50	19.17	1.2826		64 50	27.45	1.4385	
			42				33
40	19.36	1.2868		40	27.66	1.4418	
			42				33
30	19.55	1.2910		30	27.86	1.4451	
			42				32
20	19.73	1.2952		20	28.07	1.4483	
			42				32
10	19.92	1.2994		10	28.28	1.4515	
			42				32
71 00	0 20.11	1.3036		64 00	0 28.49	1.4547	
			39				32
70 50	20.30	1.3075		63 50	28.70	1.4579	
			39				32
40	20.49	1.3116		40	28.91	1.4611	
			41				32
30	20.69	1.3157		30	29.13	1.4643	
			41				31
20	20.88	1.3197		20	29.34	1.4674	
			40				32
10	21.07	1.3237		10	29.55	1.4706	
			40				30
70 00	0 21.26	1.3277		63 00	0 29.76	1.4736	
			38				32
69 50	21.45	1.3315		62 50	29.97	1.4768	
			39				31
40	21.65	1.3354		40	30.19	1.4799	
			39				30
30	22.84	1.3393		30	30.40	1.4829	
			38				31
20	22.03	1.3431		20	30.62	1.4860	
			38				30
10	22.23	1.3469		10	30.83	1.4890	
			38				31
69 00	0 22.42	1.3507		62 00	0 31.05	1.4921	
			37				31

Fahrenheit's Thermometer 50°.   Barometer 30 Inches.

Alt.	r.	Log. r.	Diff.	Alt.	r.	Log. r.	Diff.
° ′	′ ″			° ′	′ ″		
62 00	0 31.05	1.4921	31	55 00	0 40.89	1.6116	27
61 50	31.27	1.4952	30	54 50	41.14	1.6143	27
40	31.49	1.4982	31	40	41.40	1.6170	27
30	31.72	1.5013	30	30	41.65	1.6197	26
20	31.94	1.5043	30	20	41.91	1.6223	27
10	32.16	1.5073	29	10	42.16	1.6250	26
61 00	0 32.38	1.5102	31	54 00	0 42.42	1.6276	27
60 50	32.60	1.5133	29	53 50	42.68	1.6303	27
40	32.83	1.5162	30	40	42.95	1.6330	26
30	33.05	1.5192	29	30	43.21	1.6356	26
20	33.27	1.5221	29	20	43.47	1.6382	26
10	33.50	1.5250	29	10	43.74	1.6408	27
60 00	0 33.72	1.5279	29	53 00	0 44.00	1.6435	26
59 50	33.95	1.5308	29	52 50	44.27	1.6461	26
40	34.18	1.5337	29	40	44.54	1.6487	26
30	34.40	1.5366	29	30	44.80	1.6513	26
20	34.63	1.5395	29	20	45.07	1.6539	26
10	34.86	1.5423	29	10	45.34	1.6565	26
59 00	0 35 09	1.5452	29	52 00	0 45 61	1.6591	26
58 50	35.32	1.5481	29	51 50	45.89	1.6617	26
40	35.56	1.5510	28	40	46.16	1.6643	26
30	35.79	1.5538	28	30	46.44	1.6669	26
20	36.02	1.5566	28	20	46.72	1.6695	25
10	36.26	1.5594	28	10	46.99	1.6720	26
58 00	0 36.49	1.5622	28	51 00	0 47.27	1.6746	26
57 50	36.73	1.5650	28	50 50	47.56	1.6772	26
40	36.97	1.5678	29	40	47.84	1.6798	26
30	37.21	1.5707	28	30	48.13	1.6824	26
20	37.45	1.5735	27	20	48.42	1.6850	26
10	37.69	1.5762	28	10	48.70	1.6876	26
57 00	0 37.93	1.5790	28	50 00	0 48.99	1.6901	257
56 50	38.17	1.5818	27	49 50	49.28	1.69267	256
40	38.42	1.5845	28	40	49.58	1.69523	257
30	38.66	1.5873	27	30	49.87	1.69780	257
20	38.90	1.5900	27	20	50.16	1.70037	256
10	39.15	1.5927	27	10	50.46	1.70293	257
56 00	0 39.39	1.5954	27	49 00	0 50 75	1.70550	254
55 50	39.64	1.5981	28	48 50	51.06	1.70804	254
40	39.89	1.6009	27	40	51.36	1.71058	253
30	40.14	1.6036	27	30	51.66	1.71311	253
20	40.39	1.6063	27	20	51.96	1.71564	254
10	40.64	1.6090	26	10	52.27	1.71818	252
55 00	0 40.89	1.6116	27	48 00	0 52.57	1.72070	252

Fahrenheit's Thermometer 50°.   Barometer 30 Inches.

Alt.	r.	Log. r.	Diff.	Alt.	r.	Log. r.	Diff.
° ′	′ ″			° ′	′ ″		
48 00	0 52.57	1.72070	252	41 00	1 7.11	1.82678	255
47 50	52.88	1.72322	252	40 50	7.51	1.82933	255
40	53.19	1.72574	252	40	7.91	1.83188	255
30	53.50	1.72826	252	30	8.32	1.83443	255
20	53.81	1.73078	251	20	8.72	1.83698	255
10	54.12	1.73329	251	10	9.12	1.83953	255
47 00	0 54.43	1.73580	253	40 00	1 9.52	1.84208	256
46 50	54.75	1.73833	254	39 50	9.94	1.84464	257
40	55.07	1.74087	253	40	10.35	1.84721	256
30	.55.40	1.74340	253	30	10.77	1.84977	257
20	55.72	1.74593	254	20	11.19	1.85234	256
10	56.04	1.74847	253	10	11.60	1.85490	257
46 00	0 56.35	1.75100	252	39 00	1 12.02	1.85747	258
45 50	56.68	1.75352	252	38 50	12.46	1.86005	259
40	57.02	1.75604	252	40	12.89	1.86264	258
30	57.35	1.75856	252	30	13.33	1.86522	259
20	57.69	1.76108	252	20	13.77	1.86781	258
10	58.02	1.76360	251	10	14.20	1.87039	259
45 00	0 58.36	1.76611	252	38 00	1 14.64	1.87298	260
44 50	58.70	1.76863	252	37 50	15.10	1.87558	261
40	59.05	1.77115	252	40	15.55	1.87819	261
30	59.39	1.77367	252	30	16.01	1.88080	261
20	59.74	1.77619	252	20	16.47	1.88341	260
10	1 0.08	1.77871	252	10	16.92	1.88601	262
44 00	1 0.43	1.78123	252	37 00	1 17.38	1.88863	262
43 50	0.79	1.78375	253	36 50	17.86	1.89125	262
40	1.15	1.78628	252	40	18.33	1.89387	263
30	1.50	1.78880	252	30	18.81	1.89650	263
20	1.86	1.79132	253	20	19.29	1.89913	263
10	2.21	1.79385	252	10	19.76	1.90176	264
43 00	1 2.57	1.79637	253	36 00	1 20.24	1.90440	265
42 50	2.94	1.79890	253	35 50	20.74	1.90705	265
40	3.31	1.80143	253	40	21.24	1.90970	266
30	3.69	1.80396	253	30	21.75	1.91236	266
20	4.06	1.80649	253	20	22.25	1.91502	267
10	4.43	1.80902	253	10	22.75	1.91769	267
42 00	1 4.80	1.81155	254	35 00	1 23.25	1.92036	268
41 50	5.18	1.81409	254	34 50	23.78	1.92304	269
40	5.57	1.81663	253	40	24.30	1.92573	268
30	5.95	1.81916	254	30	24.83	1.92841	271
20	6.34	1.82170	254	20	25.36	1.93112	270
10	6.72	1.82424	254	10	25.88	1.93382	270
41 00	1 7.11	1.82678	255	34 00	1 26.41	1.93653	271

Fahrenheit's Thermometer 50°.   Barometer 30 Inches.

Alt.	r.	Log. r.	Diff.	Alt.	r.	Log. r.	Diff.
° ′	′ ″			° ′	′ ″		
34 00	1 26.41	1.93653	271	27 00	1 54.17	2.05754	310
33 50	26.96	1.93924	272	26 50	54.99	2.06064	312
40	27.52	1.94196	273	40	55.81	3.06376	312
30	28.07	1.94469	273	30	56.66	2.06688	315
20	28.62	1.94742	274	20	57.50	2.07003	315
10	29.18	1.95016	275	10	58.36	2.07318	317
33 00	1 29.73	1.95291	275	26 00	1 59.22	2.07635	318
32 50	30.31	1.95566	277	25 50	2 0.09	2.07953	320
40	30.90	1.95843	277	40	0.99	2.08273	321
30	31.48	1.96120	278	30	1.88	2.08594	323
20	32.06	1.96397	279	20	2.80	2.08917	324
10	32.65	1.96676	279	10	3.71	2.09241	326
32 00	1 33.23	1.96955	280	25 00	2 4.65	2.09567	327
31 50	33.85	1.97235	281	24 50	5.59	2.09894	330
40	34.46	1.97516	281	40	6.54	2.10224	330
30	35.08	1.97797	283	30	7.51	2.10554	332
20	35.70	1.98080	282	20	8.49	2.10886	334
10	36.31	1.98362	284	10	9.48	2.11220	335
31 00	1 36.93	1.98646	285	24 00	2 10.48	2.11555	337
30 50	37.58	1.98931	285	23 50	11.50	2.11892	339
40	38.24	1.99216	287	40	12.52	2.12231	340
30	38.89	1.99503	287	30	13.57	2.12571	342
20	39.54	1.99790	289	20	14.62	2.12913	345
10	40.20	2.00079	289	10	15.70	2.13258	345
30 00	1 40.85	2.00368	290	23 00	2 16.78	2.13603	348
29.50	41.52	2.00658	291	22 50	17.88	2.13951	349
40	42.21	2.00949	292	40	19.00	2.14300	352
30	42.90	2.01241	293	30	20.13	2.14652	354
20	43.59	2.01535	294	20	21.28	2.15006	355
10	44.30	2.01829	295	10	22.43	2.15361	358
29 00	1 45.01	2.02124	296	22 00	2 23.61	2.15719	359
28 50	45.73	2.02420	298	21 50	24.81	2.16078	362
40	46.46	2.02718	299	40	26.02	2.16440	364
30	47.18	2.03016	300	30	27.25	2.16804	366
20	47.93	2.03316	301	20	28.50	2.17171	368
10	48.68	2.03617	301	10	29.76	2.17539	371
28 00	1 49.44	2.03918	303	21 00	2 31.04	2.17910	373
27 50	50.21	2.04221	304	20 50	32.34	2.18283	375
40	50.99	3.04525	305	40	33.67	2.18658	378
30	51.77	2.04830	307	30	35.01	2.19036	381
20	52.57	2.05137	308	20	36.37	2.19417	383
10	53.36	2.05445	309	10	37.76	2.19800	385
27 00	1 54.17	2.05754	310	20 00	2 39.16	2.20185	388

Fahrenheit's Thermometer 50°.   Barometer 30 Inches.

Alt.	r.	Log. r.	Diff.	Alt.	r.	Log. r.	Diff.
20 00	2 39.16	2.20185	388	13 00	4 7.91	2.39430	557
19 50	40.59	2.20573	390	12 50	11.11	2.39987	563
40	42.04	2.20963	393	40	14.39	2.40550	569
30	43.52	2.21356	396	30	17.74	2.41119	576
20	45.02	2.21752	398	20	21.19	2.41695	583
10	46.53	2.22150	402	10	24.72	2.42278	589
19 00	2 48.08	2.22552	404	12 00	4 28.33	2.42867	596
18 50	49.65	2.22956	407	11 50	32.04	2.43463	603
40	51.25	2.23363	410	40	35.84	2.44066	611
30	52.87	2.23773	413	30	39.75	2.44677	618
20	54.53	2.24186	417	20	43.76	2.45295	626
10	56.21	2.24603	419	10	47.88	2.45921	635
18 00	2 57.92	2.25022	423	11 00	4 52.12	2.46556	642
17 50	59.66	2.25445	425	10 50	56.47	2.47198	650
40	3 1.43	2.25870	429	40	5 0.94	2.47848	659
30	3.23	2.26299	433	30	5.54	2.48507	669
20	5.06	2.26732	436	20	10.28	2.49176	677
10	6.93	2.27168	440	10	15.16	2.49853	688
17 00	3 8.83	2.27608	443	10 00	5 20.19	2.50541	696
16 50	10.77	2.28051	447	9 50	25.86	2.51237	707
40	12.74	2.28498	450	40	30.70	2.51944	716
30	14.75	2.28948	454	30	36.20	2.52660	727
20	16.80	2.29402	458	20	41.88	2.53387	738
10	18.88	2.29860	462	10	47.74	2.54125	749
16 00	3 21.01	2.30322	467	9 00	5 53.79	2.54874	759
15 50	23.18	2.30789	470	8 50	6 0.04	2.55635	772
40	25.39	2.31259	475	40	6.50	2.56407	785
30	27.66	2.31734	479	30	13.18	2.57192	797
20	29.95	2.32213	483	20	20.09	2.57989	811
10	32.30	2.32696	488	10	27.26	2.58800	824
15 00	3 34.70	2.33184	493	8 00	6 34.68	2.59624	838
14 50	37.16	2.33677	497	7 50	42.37	2.60462	851
40	39.65	2.34174	502	40	50.33	2.61313	866
30	42.21	2.34676	507	30	58.59	2.62179	883
20	44.82	2.35183	512	20	7 7.19	2.63062	899
10	47.48	2.35695	517	10	16.13	2.63961	914
14 00	3 50.21	2.36212	523	7 00	7 25.40	2.64875	931
13 50	53.00	2.36735	528	6 50	35.05	2.65806	949
40	55.85	2.37263	533	40	45.10	2.66755	967
30	58.76	2.37796	538	30	55.58	2.67722	986
20	4 1.74	2.38334	545	20	8 6.50	2.68708	1006
10	4.79	2.38879	551	10	17.90	2.69714	1026
13 00	4 7.91	2.39430	557	6 00	8 29.80	2.70740	1047

Fahrenheit's Thermometer 50°.   Barometer 30 Inches.

Alt.	r.	Log. r.	Diff.	Alt.	r.	Log. r.	Diff.
° ′	′ ″			° ′	′ ″		
6 00	8 29.80	2.70740	1047	3 00	14 26.04	2.93754	1608
5 50	42.24	2.71787	1069	2 50	58.71	2.95362	1654
40	55.25	2.72856	1092	40	15 33.60	2.97016	1701
30	9 8.88	2.73948	1115	30	16 10.89	2.98717	1749
20	23.16	2.75063	1139	20	50.8	3.00466	1801
10	38.12	2.76202	1165	10	17 33.6	3.02267	1855
5 00	9 53.84	2.77367	1191	2 00	18 19.6	3.04122	1909
4 50	10 10.35	2.78558	1219	1 50	19 9.0	3.06031	1967
40	27.73	2.79777	1248	40	20 2.2	3.07998	2026
30	46.03	2.81025	1277	30	59.6	3.10024	2089
20	11 5.30	2.82302	1309	20	22 1.7	3.12113	2155
10	25.66	2.83611	1340	10	23 8.9	3.14268	2221
4 00	11 47.15	2.84951	1374	1 00	24 21.8	3.16489	2290
3 50	12 9.68	2.86325	1410	0 50	25 40.9	3.18779	2361
40	33.97	2.87735	1447	40	27 7.1	3.21140	2434
30	59.51	2.89182	1484	30	28 40.8	3.23574	2509
20	13 26.61	2.90666	1523	20	30 23.2	3.26083	2584
10	55.40	2.92189	1565	10	32 15.0	3.28667	2667
3 00	14 26.04	2.93754	1608	0 00	34 17.5	3.31334	

In ordinary cases it will be sufficient to apply to the observed altitude the Mean Refraction standing against it in the adjoining column. Where greater accuracy is required, the corresponding log. *r* must be taken. In the Table of Corrections on the following page, for the Barometer and Thermometers, the proportional parts of log. *t*, for tenths of a degree of Fahrenheit, will be found in the column adjoining that of log. *t*, standing against the corresponding units of the argument. In the same manner the proportional parts of log. *β*, for hundredths of an inch, will be found standing against the corresponding tenths. These must be added or subtracted according to the sign at the top of the column. The proportional parts of log. *τ*, for tenths of a degree, will be found at the bottom. The sum of logs. *r*, *t*, *β*, and *τ*, will be the log. of the refraction, which must be subtracted from the observed altitude, or added to the observed zenith distance.

The column *Barometer* contains the Logarithms of $\frac{p}{30}$, *p* being the height of the Barometer in English Inches.

Corrections, depending on the state of the Thermometer and Barometer, to be applied to the foregoing Mean Refractions.

External Thermometer.

Th.	Log. t.	P. P.	Th.	Log. t.	P. P.
°			°		
10	0.03779	—	50	0.00000	—
1	0.03680	10	1	9.99910	9
2	0.03582	20	2	9.99820	18
3	0.03484	29	3	9.99730	27
4	0.03386	39	4	9.99640	36
5	0.03288	49	5	9.99550	45
6	0.03191	59	6	9.99460	54
7	0.03094	69	7	9.99371	63
8	0.02997	78	8	9.99282	72
9	0.02900	88	9	9.99193	81
20	0.02803		60	9.99104	
1	0.02706	10	1	9.99016	9
2	0.02609	19	2	9.98927	18
3	0.02514	29	3	9.98839	26
4	0.02418	38	4	9.98751	35
5	0.02323	48	5	9.98663	44
6	0.02227	58	6	9.98575	53
7	0.02132	67	7	9.98488	62
8	0.02037	77	8	9.98401	70
9	0.01942	86	9	9.98314	79
30	0.01848		70	9.98227	
1	0.01754	9	1	9.98140	9
2	0.01660	19	2	9.98054	17
3	0.01566	28	3	9.97967	26
4	0.01472	38	4	9.97881	34
5	0.01379	47	5	9.97795	43
6	0.01285	56	6	9.97709	52
7	0.01192	66	7	9.97623	60
8	0.01099	75	8	9.97537	69
9	0.01006	85	9	9.97452	77
40	0.00914		80	9.97367	
1	0.00822	9	1	9.97282	8
2	0.00730	18	2	9.97197	17
3	0.00638	28	3	9.97112	25
4	0.00546	37	4	9.97027	34
5	0.00455	46	5	9.96943	42
6	0.00363	55	6	9.96859	50
7	0.00272	64	7	9.96775	59
8	0.00181	74	8	9.96691	67
9	0.00090	83	9	9.96607	76
50	0.00000		90	9.96524	

Barometer.

Bar.	Log. β.	P. P.
26.6	9.94776	
7	9.94939	
8	9.95101	
9	9.95263	
27.0	9.95464	
1	9.95584	
2	9.96745	
3	9.95904	
4	9.96063	
5	9.96221	
6	9.96379	
7	9.96536	
8	9.96692	
9	9.96848	
28.0	9.97004	+
1	9.97158	15
2	9.97313	30
3	9.97466	46
4	9.97620	61
5	9.97772	76
6	9.97924	91
7	9.98076	106
8	9.98227	122
9	9.98378	137
29.0	9.98528	
1	9.98677	15
2	9.98826	29
3	9.98975	44
4	9.99123	59
5	9.99270	73
6	9.99417	88
7	9.99563	103
8	9.99709	118
9	9.99855	132
30.0	0.00000	
1	0.00145	14
2	0.00289	29
3	0.00432	43
4	0.00575	57
5	0.00718	71
6	0.00860	86
7	0.01002	100
8	0.01143	114
9	0.01284	129
31.0	0.01424	

Internal Thermometer.

Th	Log. τ.	Th.	Log. τ
°		°	
10	0.00173	50	0.00000
11	0.00169	51	9.99996
12	0.00164	52	9.99991
13	0.00160	53	9.99987
14	0.00156	54	9.99983
15	0.00151	55	9.99978
16	0.00147	56	9.99974
17	0.00143	57	9.99970
18	0.00138	58	9.99965
19	0.00134	59	9.99961
20	0.00130	60	9.99957
21	0.00126	61	9.99953
22	0.00121	62	9.99948
23	0.00117	63	9.99944
24	0.00113	64	9.99940
25	0.00108	65	9.99935
26	0.00104	66	9.99931
27	0.00100	67	9.99927
28·	0.00095	68	9.99922
29	0.00091	69	9.99918
30	0.00087	70	9.99913
31	0.00083	71	9.99909
32	0.00078	72	9.99904
33	0.00074	73	9.99900
34	0.00070	74	9.99896
35	0.00065	75	9.99891
36	0.00061	76	9.99887
37	0.00057	77	9.99883
38	0.00052	78	9.99878
39	0.00048	79	9.99874
40	0.00043	80	9.99870
41	0.00039	81	9.99866
42	0.00034	82	9.99861
43	0.00030	83	9.99857
44	0.00026	84	9.99853
45	0.00021	85	9.99848
46	0.00017	86	9.99844
47	0.00013	87	9.99840
48	0.00008	88	9.99835
49	0.00004	89	9.99831
50	0.00000	90	9.99827

P. P. to tenths of a Degree.

	.1	.2	.3	.4	.5	.6	.7	.8	.9
—	0	1	1	2	3	3	3	3	4

IV. *The Transit Instrument.*

Knowing the apparent right ascension of a star, to compute the corrections to its observed transit on account of the three principal errors of the Transit instrnment—in Azimuth, in the Inclination of the axis, and in Collimation—in order to obtain the correct clock error.

$$E = T + a \cdot \frac{\sin(L - D)}{\cos D} + b \frac{\cos(L - D)}{\cos D} + \frac{c}{\cos D} - AR$$

E = the error of the clock; *minus* when slow.

T = the observed time of transit.

L = the latitude of the place.

D = the declination of the star : *plus* when North, and *minus* when South, for the upper culminations; and *vice versa* for the lower culminations.

a = the deviation of the telescope is azimuth; *plus* when (pointing to the South) the vertical which it describes falls to the East; and *minus* when it falls to the West; and *vice versa* when pointing to the North.

b = the bias or inclination of the axis of the telescope : *plus*, when the west end of the axis is too high.

c = the error in collimation : *plus*, when the circle, described by the optical axis of the telescope (pointing to the South) falls to the East; and *minus*, when it falls to the West; and *vice versa* when pointing to the North.

AR = the Right Ascension of the star ; when the clock marks mean solar time, the mean time of transit of the object over the meridian must be substituted for AR.

1. To determine the value (*in time*) of the co-efficients *a*, *b*, *c*, in the preceding formula.

For inclination of the axis of the telescope :

$$b = \frac{d}{60} \left\{ (w + w') - (e + e') \right\}$$

Where w' and e' denote respectively the values of w and e, after *reversing* the *level*,

d = the value of each division of the level, in seconds of space.

w = the inclination of the level to the West.

e = the inclination of the level to the East.

For collimation :

$$c = \tfrac{1}{2}(t' - t) \cos D + \tfrac{1}{2}(b' - b) \cos (L - D.)$$

Where t' and b' denote respectively the values of t and b, after *reversing* the *instrument*,

D = the declination of a circumpolar star.

t = the time of the transit of the circumpolar star, deduced from an observation at a given *side* wire of the instrument.

For the deviation in azimuth :

By observations of a circumpolar star :

$$a = \frac{12^{\text{h}} - (T' - T)}{2 \cos L \, \text{tang} D} + \frac{b \cos (L - D) - b' \cos (L + D) + 2c}{2 \cos L \sin D}$$

Where T' and b' denote respectively the values of T and b, at the *lower* culmination.

Deviation in azimuth by transits of a high and low star.

$$a = \left\{ (AR' - AR) - (T' - T) \right\} \times \frac{\cos D' \cos D}{\cos L \sin (D' - D)}$$

Where T', AR', and D', denote respectively the values of T, AR, and D of the *second* star observed,

or make $\dfrac{\sin (L - D)}{\cos D}$ for the *first* star $= n$

and $\dfrac{\sin (L - D')}{\cos D'}$ for the *second* star $= n'$

then $a = \dfrac{(AR' - AR) - (T' - T)}{n' - n}$

n is negative for a star north of the zenith.

2. To find the equatorial interval of each wire from the central wire, observe the transit of a star of any declination D, then
Equatorial interval $=$ observed interval $\times \cos D$.

3. When the intervals on each side of the central wire are equal, the mean of the times of transit over each wire will denote the transit over the middle wire. But should they not be equal, a correction must be applied to obtain a correct mean.

Call I. II; IV. V, the equatorial intervals of each wire from the central wire, the instrument having, say 5 wires, then

Reduction to middle wire $= \dfrac{(I + II) - (IV + V)}{5 \cos D}$

Numerical values of $\dfrac{\sin (L - D)}{\cos D}$; $\dfrac{\cos (L - D)}{\cos D}$; $\dfrac{1}{\cos D}$ *for facilitating the method of determining the deviation of the Transit Instrument in Azimuth, by means of " high and low stars."*

For Deviation.	Star's Declination = ± D							For Level.
Star's Z D = (L — D)	1°	10°	20°	30°	40°	50°	60°	Star's Z D = (L — D)
1°	.02	.02	.02	.02	.02	.03	.03	89°
5	.08	.08	.09	.10	.11	.13	.17	85
10	.17	.17	.18	.20	.23	.27	.35	80
15	.26	.26	.27	.30	.34	.40	.52	75
20	.34	.34	.36	.39	.45	.53	.68	70
25	.42	.43	.45	.48	.55	.66	.84	65
30	.50	.51	.53	.57	.65	.77	1.00	60
35	.57	.58	.61	.66	.75	.89	1.15	55
40	.64	.65	.68	.74	.84	1.00	1.28	50
45	.71	.72	.75	.81	.92	1.10	1.41	45
50	.76	.78	.81	.88	1.00	1.19	1.53	40
55	.82	.83	.87	.94	1.07	1.27	1.64	35
60	.86	.88	.92	1.00	1.13	1.35	1.73	30
65	.90	.92	.96	1.05	1.18	1.41	1.81	25
70	.94	.95	1.00	1.08	1.23	1.46	1.88	20
75	.96	.98	1.03	1.11	1.26	1.50	1.93	15
80	.98	1.00	1.05	1.14	1.28	1.53	1.97	10
85	.99	1.01	1.06	1.15	1.30	1.55	1.99	5
89	1.00	1.01	1.06	1.15	1.30	1.55	1.99	1
For Collimation.	1.000	1.015	1.064	1.154	1.305	1.555	2.000	$= \dfrac{1}{\cos D}$

Form for record and computation.

SURVEY OF   STATION

Transits of Stars   *with*   *inch transit No.*
                  sidereal Chronometer, Hardy, No. 50.

Illuminated end of axis, west.

Date (1847) - -	October 6th.	October 6th.	October 6th.
Observer - - -	T. J. L.	T. J. L.	T. J. L.
Object - - -	π Capricorni	14 Capricorni	α Cygni.
Level - - - Value of 1 division ⎱ of scale = 7″.5 ⎰	E. 32.2 W. 33.0 E. 32 2 W. 33.0	E. 32.7 W. 32.5 E. 32.5 W. 33.3	E. 32.7 W. 32.5 E. 33.0 W. 32.5
Wires - - I II III IV V	h m s 20. 17. 33.0 17. 53.5 18. 12.7 18. 32.7 20. 18. 52.5	h m s 20. 29. 43.7 30. 02.7 30. 22.0 30. 41.7 20. 31. 00.7	h m s 20. 35. 00.0 35. 26.0 35. 52.0 36. 18.7 20. 36. 45.5
Sum - - -	184.4	110.8	142.2
Mean . - - Reduc'n to mid. wire	20. 18. 12.88 —.07	20. 30. 22.16 — .07	20. 35. 52.44 — .10
Transit on instrument for collim'n for level - - for dev'n in az'h	12.81 + .10 + .17	22.09 + .04 + .18	52.39 — .12 — .01
Transit by Chronom'r	20. 18. 13.08	20. 30. 22.31	20. 35. 52.21
A.R. of star - -	20. 18. 36.66	20. 30. 45.89	20. 36. 15.80
Error of Chronometer	23ˢ.58	23ˢ.58	23ˢ.59

Chronometer   slow of   time
at   p. m., October 6th, 1847.

Computation of the corrections a and b, in the preceding Transits.

Declination of π Capri. $= 18°\ 42$ S.

Latitude of Station $=$ L $= 43°\ 13'$ 14 Capri. $= 15°\ 29$ S.

α Cygni $= 44°\ 44$ N.

Level correction of π Capricorni.

		E. 32.2	W. 33
L $=$	$43°\ 13'$	32.2	33
D $= -$	$18°\ 42'$		
(L $-$ D) $=$	$61°\ 55'$	64.4	66

from table pape 185. $66 - 64.4 = 1.6$

$$\text{Cos} \frac{(L - D)}{\text{Cos D}} = 0.50 \qquad b = \frac{7.5}{60} \times 1.6 = 0^s.20$$

$$\text{Level correction} = b \frac{\text{Cos (L} - \text{D)}}{\text{Cos D}} = 0^s.20 \times 0.50 = 0^s.10$$

Deviation in Azimuth.

$$a = \frac{(AR' - AR) - (T' - T)}{n' - n}$$

T' and T being the times of transit corrected for level and collimation. Combining π Capri. and α Cygni.

	H. M. S.		H. M. S.
AR' $=$	20 36 15.80	T' $=$	20 35 52.22
AR $=$	20 18 36.66	T $=$	20 18 12.91
	17 39.14		17 39.31

$$n' = \frac{(\sin \text{L} - \text{D}')}{\text{Cos D}'} = \frac{\sin(-1°\ 31')}{\text{Cos } 44°\ 44'} = -0.03$$

$(AR' - AR) - (T' - T) = -0.17$

$$n = \frac{\sin(\text{L} - \text{D})}{\text{Cos D}} = \frac{\sin\ 61°\ 55'}{\text{Cos } 18°\ 42'} = +0.93$$

$$a = \frac{-0^s.17}{-0.03 - 0.93} = \frac{.17}{.96} = +0^s.18$$

Combining 14 Capri. and α Cygni , $a = + 0^s.19$.

$$\text{Correction for deviation in Azimuth of } \pi \text{ Capricorni} = a \frac{\sin(\text{L} - \text{D})}{\text{Cos D}} =$$

$$= 0^s.18 \times 0.93 = 0^s.17$$

Transit Instrument—Continued.

Rules for the direction of the deviation in azimuth, in the method of fixing a Transit Instrument in the meridian by "high and low stars."

Position of Stars.	Culmination.	Precedence.	Relative magnitude of Intervals.	Deviation.
Both south, or both north, or one south and the other north of the zenith.	Both upper	Highest or nearest to the Pole.	Obs'd greater.	W. of S.
	"		Obs'd less.	E. of S.
	"	Furthest from the Pole.	Obs'd greater.	E. of S.
	"		Obs'd less.	W. of S.
One north, and the other south of the zenith.	The northern being the lower culm'n.	Nearest to the Pole.	Obs'd less.	E. of N.
			Obs'd greater.	W. of N.
	The southern being upper culmination.	Farthest from the Pole.	Obs'd less.	W. of N.
			Obs'd greater.	E. of N.
Both north of the zenith.	One upper and one lower.	Upper.	Obs'd greater.	E. of N.
		Upper.	Obs'd less.	W. of N.
		Lower.	Obs'd greater.	W. of N.
		Lower.	Obs'd less.	E. of N.

LATITUDE.

V. *To determine the Latitude from the meridional altitude of an object whose declination is known.*

1. When the object observed is south of the zenith:

$$L = 90° + D - A = Z + D = 90° + Z - \triangle = 180° - (A + \triangle)$$

2. When the star is between the zenith and the pole:

$$L = A - \triangle = D - Z = 90° - (Z + \triangle) = A + D - 90°$$

3. When the star is between the pole and the horizon to the north:

$$L = A + \triangle = 90° + \triangle - Z = 90° + A - D = 180° - (Z + D)$$

where L = the latitude sought.

    D = the declination of the object, *minus* when south,

    $\triangle$ = its north polar distance,

    A = its meridional altitude,

    Z = its meridional zenith distance,

A and Z must be corrected for refraction ; when the sun is the object observed, A = observed altitude — (refraction — parallax) ± semi-diam.

VI. *Determination of the Latitude of a place by the method of circum-meridian altitudes.*

Reduction to meridian =

$$x = k \left\{ i \; \frac{\cos l \cos D}{\cos a} \right\} - m \, \text{tang} \, a \left\{ i \; \frac{\cos l \cos D}{\cos a} \right\}^2$$

$$k = \frac{2 \sin^2 \frac{1}{2} p}{\sin 1''} \qquad\qquad m = \frac{2 \sin^4 \frac{1}{2} p}{\sin 1''}$$

$$a = 90° + D - l$$

A = $a + x$ = the meridional altitude of the object,

a = its observed altitude—(refraction—parallax) $\pm$ semi-
    diameter.

p = its correct hour angle,

D = its declination,

l = the assumed latitude of the place,

x = the required correction in seconds.

When a *star* is the object observed and the chronometer marks *mean* time, $i = 1.005473$, log $i = 0.0023708$

When the *sun* is observed and the chronometer marks *sidereal* time, $i = 0.99455418$, log $i = 9.9976285$; and generally, when the chronometer has a large losing rate, x must be multiplied by $1 + 0.00002315 \; r$; when it has a gaining rate it must be divided by $1 + 00002315 \; r$; r being the rate in 24 hours, which must be assumed *minus* when *gaining*, and *plus* when *losing*.

The values of k and m for each value of p, are given in the following tables.

The meridian altitude $A = a + x$ for each observation; for any number of observations n,

$$\frac{a' + a'' + \dots}{n} + \frac{x' + x'' + \dots}{n} = \text{the mean, } a, \text{ of all the observed}$$

altitudes $+$ the mean, x, of all the corrections. Consequently,

1. Measure several successive altitudes of the object both before and after its meridional passage.

2. Note the times of each observation, and compute the time of the object's culmination; the differences between this and the times of each successive observation are the values of p', p'', etc., in time, for which the corresponding values of k', k'', etc., and m', m'', etc., must be taken from the tables.

3. The means k and m of these results will be introduced into the equation for the value of the correction, x, to be applied to a to obtain the meridional altitude, A, of the object.

4. If the final latitude differ much from the assumed, the computation should be repeated with the new value for l.

5. It is not necessary that the time of the object's culmination should be known with great precision, provided an equal number of altitudes be taken upon each side of the meridian, and at nearly equal distances from it.

6. The second correction, m, is seldom necessary, unless great accuracy is desired, and the object is observed more than ten minutes of time from the meridian.

Reduction to the Meridian; values of $k = \dfrac{2 \sin^2 \frac{1}{2} p}{\sin 1''}$

Sec.	0ᵐ	1ᵐ	2ᵐ	3ᵐ	4ᵐ	5ᵐ	6ᵐ	7ᵐ
	″	″	″	″	″	″	″	″
0	0.0	2.0	7.8	17.7	31.4	49.1	70.7	96.2
1	0.0	2.0	8.0	17.9	31.7	49.4	71.1	96.7
2	0.0	2.1	8.1	18.1	31.9	49.7	71.5	97.1
3	0.0	2.2	8.2	18.3	32.2	50.1	71.9	97.6
4	0.0	2.2	8.4	18.5	32.5	50.4	72.3	98.0
5	0.0	2.3	8.5	18.7	32.7	50.7	72.7	98.5
6	0.0	2.4	8.7	18.9	33.0	51.1	73.1	99.0
7	0.0	2.4	8.8	19.1	33.3	51.4	73.5	99.4
8	0.0	2.5	8.9	19.3	33.5	51.7	73.9	99.9
9	0.0	2.6	9.1	19.5	33.8	52.1	74.3	100.4
10	0.1	2.7	9.2	19.7	34.1	52.4	74.7	100.8
11	0.1	2.7	9.4	19.9	34.4	52.7	75.1	101.3
12	0.1	2.8	9.5	20.1	34.6	53.1	75.5	101.8
13	0.1	2.9	9.6	20.3	34.9	53.4	75.9	102.3
14	0.1	3.0	9.8	20.5	35.2	53.8	76.3	102.7
15	0.1	3.1	9.9	20.7	35.5	54.1	76.7	103.2
16	0.1	3.1	10.1	20.9	35.7	54.5	77.1	103.7
17	0.2	3.2	10.2	21.2	36.0	54.8	77.5	104.2
18	0.2	3.3	10.4	21.4	36.3	55.1	77.9	104.6
19	0.2	3.4	10.5	21.6	36.6	55.5	78.3	105.1
20	0.2	3.5	10.7	21.8	36.9	55.8	78.8	105.6
21	0.2	3.6	10.8	22.0	37.2	56.2	79.2	106.1
22	0.3	3.7	11.0	22.3	37.4	56.5	79.6	106.6
23	0.3	3.8	11.2	22.5	37.7	56.9	80.0	107.0
24	0.3	3.8	11.3	22.7	38.0	57.3	80.4	107.5
25	0.3	3.9	11.5	22.9	38.3	57.6	80.8	108.0
26	0.4	4.0	11.6	23.1	38.6	58.0	81.3	108.5
27	0.4	4.1	11.8	23.4	38.9	58.3	81.7	109.0
28	0.4	4.2	11.9	23.6	39.2	58.7	82.1	109.5
29	0.5	4.3	12.1	23.8	39.5	59.0	82.5	110.0

Reduction to the Meridian; values of $k = \dfrac{2 \sin^2 \frac{1}{2} p}{\sin 1''}$

Sec.	0^m	1^m	2^m	3^m	4^m	5^m	6^m	7^m
	$''$	$''$	$''$	$''$	$''$	$''$	$''$	$''$
30	0.5	4.4	12.3	24.0	39.8	59.4	83.0	110.4
31	0.5	4.5	12.4	24.3	40.1	59.8	83.4	110.9
32	0.6	4.6	12.6	24.5	40.3	60.1	83.8	111.4
33	0·6	4.7	12.8	24.7	40.6	60.5	84.2	111.9
34	0.6	4.8	12.9	25.0	40.9	60.8	84.7	112.4
35	0.7	4.9	13.1	25.2	41.2	61.2	85.1	112.9
36	0.7	5.0	13.3	25.4	41.5	61.6	85.5	113.4
37	0.7	5.1	13.4	25.7	41.8	61.9	86.0	113.9
38	0.8	5.2	13.6	25.9	42.1	62.3	86.4	114.4
39	0.8	5.3	13.8	26.2	42.5	62.7	86.8	114.9
40	0.9	5.4	14.0	26.4	42.8	63.0	87.3	115.4
41	0.9	5.6	14.1	26.6	43.1	63.4	87.7	115.9
42	1.0	5.7	14.3	26.9	43.4	63.8	88.1	116.4
43	1.0	5.8	14.5	27.1	43.7	64.2	88.6	116.9
44	1.1	5.9	14.7	27.4	44.0	64.5	89.0	117.4
45	1.1	6.0	14.8	27.6	44.3	64.9	89.5	117.9
46	1.2	6.1	15.0	27.9	44.6	65.3	89.9	118.4
47	1.2	6.2	15.2	28.1	44.9	65.7	90.3	118.9
48	1.3	6.4	15.4	28.3	45.2	66.0	90.8	119.5
49	1.3	6.5	15.6	28.6	45.5	66.4	91.2	120.0
50	1.4	6.6	15.8	28.8	45.9	66.8	91.7	120.5
51	1.4	6.7	15.9	29.1	46.2	67.2	92.1	121.0
52	1.5	6.8	16.1	29.4	46.5	67.6	92.6	121.5
53	1.5	7.0	16.3	29.6	46.8	68.0	93.0	122.0
54	1.6	7.1	16.5	29.9	47.1	68.3	93.5	122.5
55	1.6	7.2	16.7	30.1	47.5	68.7	93.9	123.1
56	1.7	7.3	16.9	30.4	47.8	69.1	94.4	123.6
57	1.8	7.5	17.1	30.6	48.1	69.5	94.8	124.1
58	1.8	7.6	17.3	30.9	48.4	69.9	95.3	124.6
59	1.9	7.7	17.5	31.1	48.8	70.3	95.7	125.1

25

Reduction to the Meridian; values of $k = \dfrac{2 \sin^2 \frac{1}{2} p}{\sin 1''}$

Sec.	8^m	9^m	10^m	11^m	12^m	13^m	14^m
	''	''	''	''	''	''	''
0	125.7	159.0	196.3	237.5	282.7	331.8	384.7
1	126.2	159.6	197.0	238.3	283.5	332.6	385.6
2	126.7	160.2	197.6	239.0	284.2	333.4	386.6
3	127.2	160.8	198.3	239.7	285.0	334.3	387.5
4	127.8	161.4	198.9	240.4	285.8	335.2	388.4
5	128.3	162.0	199.6	241.2	286.6	336.0	389.3
6	128.8	162.6	200.3	241.9	287.4	336.9	390.2
7	129.3	163.2	200.9	242.6	288.2	337.7	391.1
8	129.9	163.8	201.6	243.3	289.0	338.6	392.1
9	130.4	164.4	202.2	244.1	289.8	339.4	393.0
10	131.0	165.0	202.9	244.8	290.6	340.3	393.9
11	131.5	165.6	203.6	245.5	291.4	341.2	394.8
12	132.0	166.2	204.2	246.3	292.2	342.0	395.8
13	132.6	166.8	204.9	247.0	293.0	342.9	396.7
14	133.1	167.4	205.6	247.7	293.8	343.7	397.6
15	133.6	168.0	206.3	248.5	294.6	344.6	398.6
16	134.2	168.6	206.9	249.2	295.4	345.5	399.5
17	134.7	169.2	207.6	249.9	296.2	346.4	400.5
18	135.3	169.8	208.3	250.7	297.0	347.2	401.4
19	135.8	170.4	208.9	251.4	297.8	348.1	402.3
20	136.3	171.0	209.6	252.2	298.6	349.0	403.3
21	136.9	171.6	210.3	253.0	299.4	349.8	404.2
22	137.4	172.2	211.0	253.6	300.2	350.7	405.1
23	138.0	172.9	211.7	254.4	301.0	351.6	406.0
24	138.5	173.5	212.3	255.1	301.8	352.5	407.0
25	139.1	174.1	213.0	255.9	302.6	353.3	408.0
26	139.6	174.7	213.7	256.6	303.5	354.2	408.9
27	140.2	175.3	214.4	257.4	304.3	355.1	409.9
28	140.7	175.9	215.1	258.1	305.1	356.0	410.8
29	141.3	176.6	215.8	258.9	305.9	356.9	411.7

Reduction to the Meridian; values of $k = \dfrac{2 \sin^2 \frac{1}{2} p}{\sin 1''}$

Sec.	8ᵐ	9ᵐ	10ᵐ	11ᵐ	12ᵐ	13ᵐ	14ᵐ
	''	''	''	''	''	''	''
30	141.8	177.2	216.4	259.6	306.7	357.7	412.7
31	142.4	177.8	217.1	260.4	307.5	358.6	413.6
32	143.0	178.4	217.8	261.1	308.4	359.5	414.6
33	143.5	179.0	218.5	261.9	309.2	360.4	415.5
34	144.1	179.7	219.2	262.6	310.0	361.3	416.5
35	144.6	180.3	219.9	263.4	310.8	362.2	417.5
36	145.2	180.9	220.6	264.1	311.6	363.1	418.4
37	145.8	181.6	221.3	264.9	312.5	364.0	419.4
38	146.3	182.2	222.0	265.7	313.3	364.8	420.3
39	146.9	182.8	222.7	266.4	314.1	365.7	421.3
40	147.5	183.5	223.4	267.2	315.0	366.6	422.2
41	148.0	184.1	224.1	267.9	315.8	367.5	423.2
42	148.6	184.7	224.8	268.7	316.6	368.4	424.2
43	149.2	185.4	225.5	269.5	317.4	369.3	425.1
44	149.7	186.0	226.2	270.3	318.3	370.2	426.1
45	150.3	186.6	226.9	271.0	319.1	371.1	427.0
46	150.9	187.3	227.6	271.8	319.9	372.0	428.0
47	151.5	187.9	228.3	272.6	320.8	372.9	429.0
48	152.0	188.5	229.0	273.3	321.6	373.8	429.9
49	152.6	189.2	229.7	274.1	322.4	374.7	430.9
50	153.2	189.8	230.4	274.9	323.3	375.6	431.9
51	153.8	190.5	231.1	275.6	324.1	376.5	432.8
52	154.4	191.1	231.8	276.4	325.0	377.4	433.8
53	154.9	191.8	232.5	277.2	325.8	378.3	434.8
54	155.5	192.4	233.2	278.0	326.7	379.3	435.8
55	156.1	193.1	234.0	278.8	327.5	380.2	436.7
56	156.7	193.7	234.7	279.5	328.4	381.1	437.7
57	157.3	194.4	235.4	280.3	329.2	382.0	438.7
58	157.8	195.0	236.1	281.1	330.0	382.9	439.7
59	158.4	195.7	236.8	281.9	330.9	383.8	440.6

Reduction to the Meridian; values of $k = \dfrac{2 \sin^2 \frac{1}{2} p}{\sin 1''}$

Sec.	15ᵐ	16ᵐ	17ᵐ	18ᵐ	19ᵐ	20ᵐ	21ᵐ
	"	"	"	"	"	"	"
0	441.6	502.5	567.2	635.9	708.4	784.9	865.3
1	442.6	503.5	568.3	637.0	709.7	786.2	866.6
2	443.6	504.6	569.4	638.2	710.9	787.5	868.0
3	444.6	505.6	570.5	639.4	712.1	788.8	869.4
4	445.6	506.7	571.6	640.6	713.4	790.1	870.8
5	446.5	507.7	572.8	641.7	714.6	791.4	872.1
6	447.5	508.8	573.9	642.9	715.9	792.7	873.5
7	448.5	509.8	575.0	644.1	717.1	794.0	874.9
8	449.5	510.9	576.1	645.3	718.4	795.4	876.3
9	450.5	511.9	577.2	646.5	719.6	796.7	877.6
10	451.5	513.0	578.4	647.7	720.9	798.0	879.0
11	452.5	514.0	579.5	648.9	722.1	799.3	880.4
12	453.5	515.1	580.6	650.0	723.4	800.7	881.8
13	454.5	516.1	581.7	651.2	724.6	802.0	883.2
14	455.5	517.2	582.9	652.4	725.9	803.3	884.6
15	456.5	518.3	584.0	653.6	727.2	804.6	886.0
16	457.5	519.3	585.1	654.8	728.4	806.0	887.4
17	458.5	520.4	586.2	656.0	729.7	807.3	888.8
18	459.5	521.5	587.4	657.2	730.9	808.6	890.2
19	460.5	522.5	588.5	658.4	732.2	809.9	891.6
20	461.5	523.6	589.6	659.6	733.5	811.3	893.0
21	462.5	524.6	590.8	660.8	734.7	812.6	894.4
22	463.5	525.7	591.9	662.0	736.0	813.9	895.8
23	464.5	526.8	593.0	663.2	737.3	815.2	897.2
24	465.5	527.9	594.2	664.4	738.5	816.6	898.6
25	466.5	528.9	595.3	665.6	739.8	817.9	900.0
26	467.5	530.0	596.5	666.8	741.1	819.2	901.4
27	468.5	531.1	597.6	668.0	742.3	820.5	902.8
28	469.5	532.2	598.7	669.2	743.6	821.9	904.2
29	470.5	533.2	599.9	670.4	744.9	823.2	905.6

$$\text{Reduction to the Meridian; values of } k = \frac{2 \sin^2 \tfrac{1}{2} p}{\sin 1''}$$

Sec.	15^m	16^m	17^m	18^m	19^m	20^m	21^m
	''	''	''	''	''	''	''
30	471.5	534.3	601.0	671.6	746.2	824.6	907.0
31	472.6	535.4	602.2	672.8	747.4	825.9	908.4
32	473.6	536.5	603.3	674.1	748.7	827.3	909.8
33	474.6	537.6	604.5	675.3	750.0	828.6	911.2
34	475.6	538.7	605.6	676.5	751.3	829.9	912.6
35	476.6	539.7	606.8	677.7	752.6	831.2	914.0
36	477.6	540.8	607.9	678.9	753.8	832.6	915.5
37	478.7	541.9	609.1	680.1	755.1	833.9	916.9
38	479.7	543.0	610.2	681.3	756.4	835.3	918.3
39	480.7	544.1	611.4	682.6	757.7	836.6	919.7
40	481.7	545.2	612.5	683.8	759.0	838.0	921.1
41	482.8	546.3	613.7	685.0	760.2	839.3	922.5
42	483.8	547.4	614.8	686.2	761.5	840.7	923.9
43	484.8	548.4	616.0	687.4	762.8	842.0	925.3
44	485.8	549.5	617.2	688.7	764.1	843.4	926.8
45	486.9	550.6	618.3	689.9	765.4	844.7	928.2
46	487.9	551.7	619.5	691.1	766.7	846.1	929.6
47	488.9	552.8	620.6	692.4	768.0	847.5	931.0
48	490.0	553.9	621.8	693.6	769.3	848.9	932.4
49	491.0	555.0	623.0	694.8	770.6	850.2	933.8
50	492.0	556.1	624.1	696.0	771.9	851.6	935.2
51	493.1	557.2	625.3	697.3	773.1	852.9	936.6
52	494.1	558.3	626.5	698.5	774.5	854.3	938.1
53	495.2	559.4	627.6	699.7	775.8	855.7	939.5
54	496.2	560.5	628.8	701.0	777.1	857.1	940.9
55	497.2	561.6	630.0	702.2	778.4	858.4	942.3
56	498.3	562.7	631.2	703.5	779.7	859.8	943.8
57	499.3	563.9	632.3	704.7	781.0	861.1	945.2
58	500.3	565.0	633.5	705.9	782.3	862.5	946.6
59	501.4	566.1	634.7	707.1	783.6	863.9	948.1

Reduction to the Meridian; values of $k = \dfrac{2 \sin^2 \frac{1}{2} p}{\sin 1''}$

Sec.	22^m	23^m	24^m	Sec.	22^m	23^m	24^m
	''	''	''		''	''	''
0	949.6	1037.8	1129.9	30	993.2	1083.3	1177.5
1	951.0	1039.3	1131.4	31	994.7	1084.8	1179.1
2	952.4	1040.8	1133.0	32	996.2	1086.4	1180.7
3	953.8	1042.3	1134.6	33	997.6	1087.9	1182.3
4	955.3	1043.8	1136.2	34	999.1	1089.5	1183.9
5	956.7	1045.3	1137.8	35	1000.6	1091.0	1185.5
6	958.2	1046.8	1139.3	36	1002.1	1092.6	1187.1
7	959.6	1048.3	1140.9	37	1003.5	1094.1	1188.7
8	961.1	1049.8	1142.5	38	1005.0	1095.7	1190.3
9	962.5	1051.3	1144.0	39	1006.5	1097.2	1191.9
10	963.9	1052.8	1145.6	40	1008.0	1098.8	1193.5
11	965.4	1054.3	1147.2	41	1009.4	1100.3	1195.1
12	966.9	1055.9	1148.8	42	1010.9	1101.9	1196.7
13	968.3	1057.4	1150.4	43	1012.4	1103.4	1198.3
14	969.8	1058.9	1152.0	44	1013.9	1105.0	1199.9
15	971.2	1060.4	1153.6	45	1015.4	1106.5	1201.5
16	972.7	1062.0	1155.2	46	1016.9	1108.1	1203.1
17	974.1	1063.5	1156.8	47	1018.4	1109.6	1204.7
18	975.5	1065.0	1158.3	48	1019.9	1111.2	1206.4
19	977.0	1066.5	1159.9	49	1021.4	1112.7	1208.0
20	978.5	1068.1	1161.5	50	1022.8	1114.3	1209.6
21	979.9	1069.6	1163.1	51	1024.3	1115.8	1211.2
22	981.4	1071.1	1164.7	52	1025.8	1117.4	1212.9
23	982.9	1072.6	1166.3	53	1027.3	1118.9	1214.5
24	984.4	1074.2	1167.9	54	1028.8	1120.5	1216.1
25	985.8	1075.7	1169.5	55	1030.3	1122.0	1217.7
26	987.3	1077.2	1171.1	56	1031.8	1123.6	1219.4
27	988.8	1078.7	1172.7	57	1033.3	1125.1	1221.0
28	990.3	1080.3	1174.3	58	1034.8	1126.7	1222.6
29	991.8	1081.8	1175.9	59	1036.3	1128.3	1224.2

Second part of the Reduction to the Meridian;

$$values\ of\ m = \frac{2 \sin^4 \frac{1}{2} p}{\sin 1''}$$

Minutes.	0^s	10^s	20^s	30^s	40^s	50^s
	$''$	$'$	$''$	$''$	$''$	$''$
5	0.01	0.01	0.01	0.01	0.01	0.01
6	0.01	0.01	0.01	0.02	0.02	0.02
7	0.02	0.02	0.03	0.03	0.03	0.04
8	0.04	0.04	0.05	0.05	0.05	0.06
9	0.06	0.06	0.08	0.08	0.08	0.09
10	0.09	0.10	0.11	0.11	0.12	0.13
11	0.14	0.15	0.15	0.16	0.17	0.18
12	0.19	0.20	0.22	0.23	0.24	0.25
13	0.27	0.28	0.30	0.31	0.33	0.34
14	0.36	0.38	0.39	0.41	0.43	0.45
15	0.47	0.49	0.52	0.54	0.56	0.59
16	0.61	0.64	0.67	0.69	0.72	0.75
17	0.78	0.81	0.84	0.88	0.91	0.95
18	0.98	1.02	1.06	1.09	1.13	1.18
19	1.22	1.26	1.30	1.35	1.40	1.44
20	1.49	1.54	1.60	1.65	1.70	1.76
21	1.82	1.87	1.93	1.99	2.06	2.12
22	2.19	2.25	2.32	2.39	2.46	2.54
23	2.61	2.69	2.77	2.85	2.93	3.01
24	3.10	3.18	3.27	3.36	3.45	3.55
25	3.64	3.74	3.84	3.94	4.05	4.15
26	4.26	4.37	4.48	4.60	4.72	4.83
27	4.96	5.08	5.20	5.33	5.46	5.60
28	5.73	5.87	6.01	6.15	6.30	6.44
29	6.59	6.75	6.90	7.06	7.22	7.38
30	7.55	7.72	7.89	8.06	8.24	8.42
31	8.61	8.79	8.98	9.17	9.37	9.57
32	9.77	9.97	10.18	10.39	10.61	10.82
33	11.04	11.27	11.50	11.73	11.96	12.20
34	12.44	12.69	12.94	13.19	13.45	13.71
35	13.97	14.24	14.51	14.78	15.06	15.35

FORM FOR RECORD

SURVEY OF   DETERMINATION OF THE LATITUDE,
                                    North and South

DATE AND STATION.—1843, *October* 13—*Mouth of the Big Black river,*

NAME OF STAR, γ *Pegasi, South of the Zenith.*

INSTRUMENTS. { Sextant No. 2197, by *Troughton & Simms*, and
{ *Mean Solar* Chronometer No. 76, by *Charles*

No. for reference.	Times of observation by Chronometer.	MERIDIAN DISTANCES, $= p$.		$\dfrac{2\,\mathrm{Sin}^2\,\frac{1}{2}p}{\mathrm{Sin}\,1''} = k$	$\dfrac{\mathrm{Cos}\,l\,.\,\mathrm{cos\,D.}}{\mathrm{Co\ sine\ a_{i}}}$	Reduction to the meridian (in arc) $= x$.
		In mean Solar time.	In Sidereal time.			
	h m s	m s	m s	''		' ''
1	10 18 40.4	9 44.2	9 45.8	187.3		3 49.8
2	19 44.4	8 40.2	8 41.6	148.3		2 51.9
3	20 48	7 36.5	7 37.7	114.2		2 27.3
4	21 46.4	6 38.2	6 39.3	86.9		1 47.6
5	22 44.4	5 40.2	5 41.1	63.4		1 17.8
6	23 54	4 30.5	4 31.2	40.1		0 49.2
7	25 12	3 12.6	3 13.1	20.3	Constant multiple 1.227	0 24.9
8	26 46	1 38.6	1 38.8	5.2		0 06.3
9	28 16.4	0 08.2	0 08.2	0.0		0 00.0
10	29 42	1 17.4	1 17.6	3.2		0 03.9
11	31 42	3 17.4	3 17.9	21.4		0 26.2
12	32 54.4	4 29.8	4 30.5	40.0		0 49
13	34 18	5 53.4	5 54.3	68.5		1 24
14	36 14.2	7 49.6	7 50.9	123.5		2 31.5
15	38 32.2	10 07.6	10 09.2	202.3		4 08.2
16	40 06	11 41.4	11 43.3	269.9		5 31.1

Observer, *Major J. D. Graham.*
Computer,   *do.   de.*

AND COMPUTATION.

from observed double circum-meridian altitudes of Stars,
of the Zenith.

a tributary to the river St. John, Maine.

artificial horizon of Mercury.

Young.

Obser'd double circum-meridian altitudes of Star.	True circum-meridian altitude of Star, as corrected for refraction and errors of instrument, $= a.$	True meridian altitudes deduced, $= (a + x) = A$	Latitude, deduced from each observation $= L = (90° + D - A)$
° ′ ″	° ′ ″	° ′ ″	° ′ ″
114 34 15	57 18 38.5	57 22 28.3	46 56 42.55
36 15	57 19 38.5	57 22 30.4	56 40.45
37 10	57 20 06	57 22 33.3	56 37.55
38 10	57 20 36	57 22 23.6	56 47.25
39 30	57 21 16	57 22 33.8	56 37.05
40 30	57 21 46	57 22 35.2	56 35.65
41 05	57 22 03.5	57 22 28.4	56 42.45
41 50	57 22 26	57 22 32.3	56 34.55
41 50	57 22 26	57 22 26	56 40.85
41 50	57 22 26	57 22 29.9	56 36.95
41 00	57 22 01	57 22 27.2	56 39.65
39 45	57 21 23.5	57 22 12.5	56 58.35
38 40	57 20 51	57 22 15	56 55.85
36 30	57 19 46	57 22 17.5	56 53.55
33 20	57 18 11	57 22 19.2	56 51.85
30 50	57 16 56	57 22 27.1	46 56 39.75

LATITUDE—Deduced from a mean of 16 altitudes of
      Star γ *Pegasi* . . . . . 46° 56′ 43″.4
    Deduced from a mean of 10 altitudes of
      Star γ *Cephei*, observed this night with
      same Sextant 46 57 10.7

Mean; or Latitude adopted . . . . . 46° 56′ 57″

26

D = apparent declination of Star 14° 19′ 10″.85 N.   Log cos   9.98629
l = approximate Lat. of place 46° 57′   .   .   Log cos   9.83418

                                             Sum   19.82048
a = approximate merid. alt. of Star 57° 22′ 10″   Log cos   9.73176

$$\frac{\text{Cos } l \text{ cos D}}{\text{Cos a}} = \text{constant multiple} = 1.227 \quad . \quad . \text{ Log} \qquad 0.08872$$

Refraction (Ther. 28°, Bar. 29.14 in.) for mean obs'd alts.   — 39″
Index error of Sextant . . . .   + 2′ 40″
*Error of excentricity, &c., of Sextant   .   .   .   .   .   + 1′ 40″

                                                     h   m   s
Apparent AR. of of the Star γ Pegasi 0 05 14.09
Sidereal time at mean noon at this station 13 26 20.83

Sidereal interval from mean noon, of Star's culmination 10 38 53.16
Retardation of mean on Sidereal time   .   .   .   .   .   — 1 44.96

Mean time of culmination of Star γ Pegasi 10 37 08.2
Chronometer (C. Y. 76) *slow* of mean time at time of
    observation   .   .   .   . —08 43.6

Time by Chronometer of culmination of Star γ Pegasi 10 28 24.6

~~~~~~~~~~~~~~~

On this night, Oct. 13, 1843, Major Graham obtained for
    the Latitude of this station, from 75 observations on 5
    stars South of the zenith, combined with 21 observa-
    tions on γ Cephei and Polaris, to the North .   .   . 46° 56′ 56.″3
On the night of Oct. 24, by 43 observations on 4 south-
    ern stars, combined with 2 observations on γ Cephei,
    the Latitude deduced was 46 56 57.2
On Sept. 17, 1844, 66 observations on N. and S. stars
    gave for the Latitude of this station 46 56 60.4

*Note.—The error of excentricity is approximately ascertained by comparing
Latitudes, well determined, by observations on N. and S. stars, with that which
will result from N. or S. stars individually of various meridional altitudes. It varies with the altitudes observed. That is to say, it is different for different parts
of the limb of the instrument.
~~~~~~~~~~~~~~~

VII. *To determine the Latitude by an altitude of a star near the pole, at any hour.*

$$L = A - (\Delta \cos p) + \alpha (\Delta \sin p)^2 \tang A - \beta (\Delta \sin p)^2 (\Delta \cos p)$$

where A = the observed altitude, corrected for refraction, etc.

Δ = the polar distance of the star, in seconds of arc.

$\alpha = \frac{1}{2} \sin 1''$, $\log \alpha = 4.3845449$,

$\beta = \frac{1}{3} \sin^2 1''$, $\log \beta = 8.89403$,

p = the hour angle of the star.

$\pm p$ = sidereal time $-$ AR $\ast$ = solar time $+$ AR$\odot$ $-$ AR $\ast$

p is *plus* when the star is west, and *minus* when it is east of the meridian.

The sign of $\cos p$ should also be attended to, for when p is greater than 6^{hrs} or $90°$, the cosine is negative, and the second and fourth terms change the sign *minus* to *plus*.

The fourth term may be generally omitted; its greatest value being only $0''.55$.

This formula is only applicable to stars within a very few degrees of the pole.

For other circumpolar stars,

$$\tang x = \tang \Delta \cos p$$

$$\sin y = \frac{\cos x \sin A}{\cos \Delta}$$

$$L = y \mp x$$

In which the upper sign is used when the star is above the pole, the under when below the pole.

FORM FOR

SURVEY OF   DETERMINATION OF THE

DATE AND STATION.—*1843, September 6—Woodstock, New Brunswick,*

NAME OF STAR.—*Polaris (α Ursæ Minoris,) observed on, between four and*

INSTRUMENTS.   { Sextant No. 2197, by *Troughton & Simms,* and arti-

                  { *Mean Solar* Chronometer No. 2440, by *Parkinson &*

No. for ref.	Times of observation by *Mean Solar* Chronometer *No.* 2440.	True Sidereal times of observation.	MERIDIAN DISTANCES.		$-\Delta \cos p.$
			In Sid'l time $= p.$	In arc $= p.$	
	h. m. s.	*h. m. s.*	*h. m. s.*	° ′ ″	′ ″
1	1 33 02.5	20 05 34.1	4 58 23.2	74 35 48	—24 18.1
2	1 34 28	20 06 59.8	4 56 57.5	74 14 22.5	—24 54.5
3	1 35 42.7	20 08 14.7	4 55 42.6	73 55 39	—25 19.8
4	1 36 38.2	20 09 10.4	4 54 46.9	73 41 43.5	—25 41.4
5	1 39 07.5	20 11 40.1	4 52 17.2	73 04 18	—26 34.7
6	1 41 11.2	20 13 44.1	4 50 13.2	72 33 22.5	—27 27.1
7	1 44 28.2	20 17 01.7	4 46 55.6	71 43 54	—28 40.8

Observer, *Major J. D. Graham.*

Computer,   *Do.*

RECORD AND COMPUTATION.

LATITUDE, *from observed double altitudes of* Polaris.

(*Grover's Inn.*)
five hours before its upper meridian passage.
ficial horizon of Mercury.
Frodsham.

$+\ a\ (\Delta\ sin\ p)^2.$ $tang\ A.$	$-\beta\ (\Delta\ sin\ p)^2$ $.\ (\Delta\ cos\ p.)$	Obser'd double alts. of *Polaris* out of the Meridian.	True altitudes of Star, as corrected for refraction and errors of instrument, $= A.$	Latitude deduced from each observation $= L.$
′ ″	′	° ′ ″	° ′ ″	° ′ ″
+ 1 11.63	— 0.32	93 01.30	46 31 58.6	46 08 51.8
+ 1 11.41	— 0.33	93 02.45	46 32 36	46 08 52.6
+ 1 11.20	— 0.33	93 03.50	46 33 08.6	46 08 59.7
+ 1 11.04	— 0.33	93 04.40	46 33 33.6	46 09 02.9
+ 1 10.63	— 0.34	93 06.15	46 34 21	46 08 56.6
+ 1 10.28	— 0.35	93 08.20	46 35 23.5	46 09 06.3
+ 1 09.68	— 0.37	93 10.50	46 36 38.5	46 09 07

LATITUDE—deduced from a mean of 7 altitudes of Star
Polaris . . . . } 46° 08′ 59″.4

Apparent declination of Star 88° 28′ 30″.5.
Apt. N. P. D. of Star = 1° 31′ 29″.5 = 5489″.5 = Δ
Refraction (Ther. 57° — Bar. 30.013 inches) . . . . . — 55″.4
Index error of Sextant . . . . + 2′ 50″
Errors of excentricity &c. of Sextant . . . . . . + 1′ 28″

                  _h. m. s._
Apparent AR. of the Star _Polaris_ (_α Ursæ Minoris_) . . 1 03 57.3
Sidereal time at mean noon at this station 11 00 27.1

Sidereal interval from mean noon, of Star's culmination . 14 03 30.2
Retardation of mean on Sidereal time . . . . — 2 18.2

Mean time of culmination of Star _Polaris_ 14 01 12
Chron. No. 2440, fast of mean time at time of observation 4 29 24.8

Time by Chronometer of culmination of Star _Polaris_ . . 6 30 36.8

 The reduction of the mean time of observation to sidereal time, in the preceding example, might have been omitted by using table of _AR. in arc into mean time_, pages 152, &c. Thus—(1st observation)

Mean time of observation 1ʰ 33ᵐ 02ˢ.5
Mean time culmination of Polaris . . . . 6 30 36 .8

Hour angle, _p_, in intervals of mean time . . . . . 4 57 34 .3

Sidereal equivalents, in arc . . . . 4ʰ = 60° 09′ 51″.39
            57ᵐ = 14 17 20 .45
            34ˢ =   8 31 .40
            .3 =   4 .51

  _p_, in arc . . . . = 74° 35′ 47″.75

 FORM FOR COMPUTATION—(1st observation)

1st term.		2d term.		3d term.
log cos p (+)=	9.4242480	sin p = 9.98411		
" Δ =	3.7395327	Δ = 3.73953		
Δ cos p =	3.1637807	Δ sin p = 3.72364		. . . = 3.16378
=	1458″.1			
1st term =	—24′ 18″.1	$(\Delta$ sin $p)^2$ = 7.44728		. . . = 7.44728
		log α = 4.38454		log β = 8.89403
A =	46° 31 58 .6	tang A = 0.02325		
				9.50509
	46 07 40 .5	1.85507		3d t'm =—0″.32
2d term =	+ 1 11 .63	= 71″.63		
		2d term =+1′ 11″.63		
	46 08 52 .13			
3d term =	— 0 .32			
Latitude =	46°08′ 51″.81			

VIII. *Determination of the Latitude by transits over the prime vertical.*

Suppose a Transit instrument so placed, that the transit axis is on the meridian, or very nearly so, and that the axis is horizontal, and the collimation nothing :

1. Call the time T, at which a star, whose declination is D, passes the middle wire of the instrument on the eastern side of the meridian, the clock correction to reduce the observed time to the true E, and the right ascension of the star AR; and let T$'$ and E$'$ denote the corresponding quantities for the western transit. Then the two-hour angles, in sidereal time, will be, the eastern negative,

$$ t = \text{T} + \text{E} - \text{AR} , \qquad t' = \text{T}' + \text{E}' - \text{AR} . $$

Let the unknown Latitude of the place be L, and the Azimuth of the line of collimation, a. The spherical triangle, formed by great circles connecting the Zenith, the Pole, and the place of the Star, gives the following relations :

$$ \cot a = \frac{\cos t \cos \text{D} \sin \text{L} - \sin \text{D} \cos \text{L}}{\cos \text{D} \sin t} $$

$$ = \frac{\cos t' \cos \text{D} \sin \text{L} - \sin \text{D} \cos \text{L}}{\cos \text{D} \sin t'} $$

Whence,

$$ \text{tang L} = \text{tang D} \, \frac{\cos \frac{1}{2} (t' + t)}{\cos \frac{1}{2} (t' - t)} $$

If the instrument is very nearly on the prime vertical, $\cos \frac{1}{2} (t' + t) = \cos 0° = 1$, and

$$ \text{tang L} = \text{tang D sec.} \, \tfrac{1}{2} (t' - t) $$

for the passage over the middle wire of the instrument.

2. Call the time of passage of the Star, from a side wire to the middle wire, τ.

Let the distance, in arc, of one of the lateral wires from the middle wire, measured on a great circle, be $15 f$; f being the equatorial interval of the wire, in time.

Then, to reduce the transit over a side wire, to the centre wire,

$$\tau = \frac{f}{\left\{ \sin (L + D) \cdot \sin (L - D) \pm \tfrac{15}{2} f \right\}^{\frac{1}{2}}}$$

The upper sign of the term $\pm \tfrac{15}{2} f$, is to be used for wires crossed by the Star earlier than the middle wire in the eastern transit, and later in the western transit, and the lower sign in the opposite cases. An approximate latitude may be used for L.

3. Should the optical axis not coincide with the middle wire, substitute $f \pm c$, for f in the above, according as the error of collimation c, lies on the same or opposite sides of f.

4. The preceding formula gives the latitude on the supposition that the axis of the instrument is parallel to the horizon. If the instrument is on the prime vertical, but the north end of the axis is, for instance, n seconds too high, the axis is parallel to the horizon of a place whose latitude is n seconds less than where the instrument is placed, and the true latitude is, therefore,

$$L + n$$

5. But should the instrument not be on the prime vertical, the true latitude becomes

$$L + n \sin a$$

a being the Azimuth of the centre wire of the telescope, supposed in collimation.

This may be found from the time elapsed between the E. and W. transits of the same star, thus :

$$\cot u = \tan \tfrac{1}{2} \, (t' - t) \sin D .$$

$$\sin a = \cos D \, \frac{\sin u}{\cos L} .$$

a is taken between 0° and 90° when the north end of the transit axis is between the north and west, and between 90° and 180° when the same end is between the north and east.

If *n* is called *plus* when the north end of the axis is too high, and *vice versa*, the signs of the corrections are indicated by those of the quantities resulting from the formula.

When *a* is nearly 90°, the correction is exceedingly small ; so that, when the instrument is placed nearly east and west, we may proceed in all the computations as if it were exactly so.

6. The instrument should be set up in the firmest manner. A change of Azimuth between the east and west transits of a Star will affect the result much less than an equal change of level.

It is better, in order to obtain a close result in the shortest time, to observe several Stars on the same evening, and between the first and last observations to determine with the level the inclination of the axis several times, and then to interpolate for transits between the times of observation of the level. It is of course understood that the changes of inclination must be small, which will be the case if the instrument is properly placed.

27

7. In order to point the telescope rightly, the hour angles and zenith distances of the Stars to be observed must be computed for the time of transit.

When the telescope is on the prime vertical, calling p the hour angle, and z, the zenith distance of the Star, then

$$\cos p = \text{tang } D \cot L$$

$$\cos z = \frac{\sin D}{\sin L}$$

An allowance must be made for the time of crossing the first wire, and for change of zenith distance from the first to the middle wire.

8. To correct, for errors of Collimation, irregularity in the pivots, etc., the instrument may be reversed between the transits over each vertical; i. e., the wires on one side of the centre wire are observed, the instrument reversed in its Y's, and the transit over the same wires continued, but in an inverse order. So that, in each vertical the same wire is at one time as far north as it is at another south of the optical axis.

Then let L = the latitude sought,

    D = the apparent declination of the Star,

    t = the hour angle, illuminated axis *north*,

      = $\frac{1}{2}$ diff. of sidereal time of transit over the same wire, for same position of axis.

    t = hour angle, illuminated axis *south*,

$$\text{tang } L = \frac{\text{tang } D}{\cos \frac{1}{2} (t' + t) \cdot \cos \frac{1}{2} (t' - t)}$$

IX. *To determine the Latitude of a place, by observing the difference of the meridional zenith distances of two Stars on opposite sides of the zenith,* with the zenith and equal altitude telescope.

Compute an approximate latitude by the formula.

$$L = \tfrac{1}{2} \left\{ 180° - (\Delta + \Delta') \right\} + \tfrac{1}{2}(z - z')$$

where Δ and Δ' are the polar distances of the south and north Stars respectively, and $(z - z')$ the quantity measured by the micrometer. Then,

1. The *correction for level* is applied by *adding* the angle which the vertical axis of the instrument makes with the zenith, when its inclination is *southward,* or subtracting it when to the northward. This correction is found by multiplying the value of 1 division of the level scale, in arc, by one-half the mean change, in level divisions, which any one end of the bubble undergoes by reversing the instrument on the meridian; or, if o and e, o' and e' denote the readings of the object and eye-ends of the bubble, for south and north stars respectively; corrections for level $= \tfrac{1}{4}(o' - e') - \tfrac{1}{4}(o - e) \times$ the value of 1 division of the level scale in arc.

2. The correction for *error of meridional position* of the central vertical wire, is found by computing the usual "reduction to the meridian" for each star; then the difference between the reductions for the northern and southern stars is taken, and one-half that difference added or subtracted, according as the reduction for the *northern* star is *greater* or less than that for the southern; or,

$$\text{correction for position} = \frac{m' - m}{2}$$

m being the reduction for stars south, and m' for stars north of the zenith.

3. The *correction for Refraction* is applied similarly to this last, but with a contrary sign; or,

$$\text{correction for refraction} = \frac{r - r'}{2}$$

$(r - r')$ being small, no note need be taken of the state of the barometer and thermometer at the time of observation. It is sufficient to use the actual tabular quantities.

Including all the corrections, the general expression for Latitude will be

$$L = \frac{180 - (\Delta + \Delta')}{2} + \frac{(z - z')}{2} \cdot a + \frac{(o' + e) - (o + e')}{4} \cdot b$$
$$+ \frac{(m' - m)}{2} + \frac{(r - r')}{2}.$$

a and b being the arc values of 1 division of the micrometer and level scale respectively.

4. Should the Star be observed on one side or the other of the central wire, the reduction to the meridian becomes

$$m = \frac{225}{4} \sin 1'' \cdot \sin 2\,\Delta \cdot p^2$$

$$= [6.4356974] \sin 2\,\Delta \cdot p^2$$

p being the hour angle of the Star in seconds of *time*. Sine $2\,\Delta$ is negative when the Star is south of the equator or sub-polo.

5. To find the value a, of 1 division of the microme-ter, note the time by chronometer of the transit of Polaris over the moveable wire placed vertically, and set succes-sively to, say, every hundredth division of its scale. Then let x be the angular distance *from the meridian* at which any reading of the screw was had; p, the hour angle of the Star at the same instant, and Δ its polar dis-tance,

$$\sin x = \sin \Delta \sin p.$$

The value of x is computed for each reading, and the differences of these values, divided by the differences of the corresponding micrometer readings, give values for the screw.

6. The value b, of 1 division of the level scale will be best found by using, in conjunction with the micrometer, a distant point as a mark, or the central wire of another instrument used as a collimator; for the space above or below the mark, passed over by the horizontal wire of the micrometer, during the bubble's run over the scale, as the telescope's elevation is gradually altered, may afterwards be measured by the micrometer screw.

7. To correct, as much as possible, an erroneous deter-mination of the value of the micrometer screw, select stars for observation such, if practicable, that the greatest Z. D. of a pair will belong as often to the N. Star as to the S. Star; for if the Z. D. of the N. Star is the greatest, the observed quantity is subtractive; if least, additive. For, as a general rule, the error of *latitude*, arising from an erroneous value to the micrometer screw, will be the least when in a set of stars,

$$\Sigma z - \Sigma z' = 0.$$

FORM FOR RECORD AND COMPUTATION.

Latitude, Station   *Observations with Zenith Telescope*   *by*

Date.	No. * B.A.C.	N. or S.	Micrometers A.	Micrometers B.	Diff. Z. D. By microm.	Diff. Z. D. In arc.	Polar distances.	Approximate Latitude.	Levels O.	Levels E.	State of Level.	Meridian distances.	Reduction to meridian.	Mean refraction.	Corrections Level.	Corrections Merid'n.	Corrections Refraction.	Latitude.
						''	o ' ''					s.	''	''	''	''	''	o ' ''
1846. Sept. 20	6640	N.	....	1388.3	....	....	32 38 10.29	42° 36'	36.2	49.0	— 6.4	22.0	0 40	15.38				
	6690	S.	276.5	....	1664.8	752.01	62 21 06.28	....	35.0	49.8	+ 7.4	25.0	0.85	15.63				
							94 59 17.07											
							85 00 42.93											
							12 32.01											
							85 13 14.94	37''.47	....	....	....	....	....	....	+ 0.65	— 0.22	+ 3.10	42 36 38.03
Sept. 20	6862	N.	....	255.5	....	....	29 47 01.20	....	41.7	42.0	— 0.1	39.0	1.03	18.52				
	6966	S.	....	1282.7	1027.2	464.00	64 51 56.23	....	41.2	43.3	+ 1.0	24.0	0.73	18.38				
							94 38 57.43											
							85 21 02.57											
							7 44.00											
							85 13 18.57	39''.28	....	....	....	....	....	....	+ 0.57	+ 0.15	— 0.07	42 36 39.93

X. *Knowing the time and the latitude of the place, to find the Azimuth of the Sun or a Star.*

$$\tan \tfrac{1}{2}\,(A + S) = \cot \tfrac{1}{2}\,p\,\frac{\cos \tfrac{1}{2}\,(\Delta - \lambda)}{\cos \tfrac{1}{2}\,(\Delta + \lambda)},$$

$$\tan \tfrac{1}{2}\,(A - S) = \cot \tfrac{1}{2}\,p\,\frac{\sin \tfrac{1}{2}\,(\Delta - \lambda)}{\sin \tfrac{1}{2}\,(\Delta + \lambda)},$$

$$A = \tfrac{1}{2}\,(A + S) \mp \tfrac{1}{2}\,(A - S)$$

the *upper* or *negative* sign is used when λ is greater than Δ.

Where

A = the azimuth counted from the *north*, which must be subtracted from 180° if counted from the south.

S = the angle at the star, called the angle of variation.

λ = the co-latitude of the place.

Δ = the north polar distance of the sun or star.

p = the hour angle at the pole.

XI. *Without the use of a chronometer, by observing the altitude of the sun or star at the same instant with the observation of the azimuth.*

Let Z = the zenith distance, corrected for refraction, parallax, and semidiameter.

$$\cos^2 \tfrac{1}{2}\,A = \frac{\sin k \,.\, \sin (k - \Delta)}{\sin Z \sin \lambda}$$

$$2\,k = Z + \Delta + \lambda$$

XII. *To find the* amplitude *of a celestial object at its rising or setting; by amplitude is meant the complement of the azimuth, or distance from the east or west points of the horizon.*

This is a particular case of the preceding problem. When the object appears to be in the horizon, its zenith distance, instead of being 90°, is, on account of refraction and parallax, $90° + k$.

Where k = hor. refraction — hor. parallax
= 33′ 45″ — hor. parallax.

For stars, the hor. par. = 0 and k = 90° 33′ 45″, for the sun, k = 33′ 45″ — 8″ 6 and k = 90° 33′36″.4; the mean refraction and mean hor. par. are here used as these observations are not susceptible of a great degree of accuracy.

XIII. *To find the true meridian by the method of equal altitudes of the Sun.*

The instrument remaining stationary, observe the readings of the horizontal limb when the altitude of the Sun's centre, or of either limb, is the same in the forenoon and afternoon.

Then, the correction to the mean of these two readings for the change in the sun's declination in the interval, is

$$c = \frac{\frac{1}{2}(D - D')}{\cos L \cdot \sin \frac{1}{2}(t - t')}$$

where

$D - D'$ = the change in the sun's declination in the interval of the observations,

$(t - t')$ = this interval of time, expressed in arc

L = the latitude of the place.

XIV. *To find the azimuth of Polaris at its greatest eastern or western elongation.*

Cos $p =$ tang Δ cot $\lambda =$ cot D tang L $=$ tang L tang Δ.
Cos L sin A $=$ sin $\Delta =$ cos D, where,

$p =$ the hour angle of the Star,　A $=$ the required azimuth,
D $=$ its declination,　　　　　L $=$ the lat. of the place,
$\Delta =$ its polar distance,　　　$\lambda =$ the co-latitude.

The first equations give the hour angle of the Star at its greatest elongation; hence the sidereal time of elongation.

The second, the azimuth of the Star at its greatest elongation.

The azimuth at *any* hour angle is found by the methods X and XI, or by the formula

$$A \text{ (in seconds)} = \frac{\sin p}{\cos L} \left\{ \Delta + \Delta^2 \sin 1'' \cos p \text{ tang L} \right\}$$

The most approved method is to observe a series of azimuths of Polaris *about* the elongation, say for not more than 30 minutes before and after, and to reduce them to the elongation; to do this, compute from the known latitude, the azimuth of the Star at its greatest elongation $=$ A, and call the sidereal time from elongation t; the correction to the azimuth will be,

$$c = (112.5) \; t^2 \sin 1'' \text{ tang A}$$
$$\log (112.5) \sin 1'' = 6.7367274.$$

The quantities found in the tables for "reduction to the meridian" $\left(2 \dfrac{\sin^2 \frac{1}{2} p}{\sin 1''} \right)$ correspond very nearly to $(112.5) \; t^2 \sin 1''$, when t does not exceed $15'$; so, by entering the table with the time from elongation, and multiplying the tabular quantities by tang A, we obtain the

28

required correction in seconds of arc.   This will be found a convenient substitute for the more rigorous method.

In these observations, the optical axis of the telescope of the theodolite must be made to describe a truly vertical plane.

If the axis of the telescope is not horizontal, the correction to the azimuth will be

$$\pm \frac{d}{4}\Big[(w + w') - (e + e')\Big] \text{ tang } \ast\text{'s altitude}$$

where $d =$ the value of one division of the level scale,

$\quad w =$ the inclination of the level to the *west*,

$\quad e =$ the inclination of the level to the *east*,

w' and e', the same values after *reversing* the level.

XV. *Correction for* Run *in Reading Microscopes.*

As it is difficult to adjust the microscopes so that five revolutions of the micrometer screw shall carry the wire exactly over one of the five-minute spaces on the limb of the instrument, (if it be so graduated,) it is preferred to observe the number of revolutions and the part of a revolution made by the screw while the wire passes over the space; then

Let $m =$ the mean of *first readings*, that is, the readings obtained by turning the screw in the direction of increasing numbers from zero of the comb.

$\quad m' =$ the mean of second, or reverse, readings.

Then, (mean) Run $= r = m - m' + 300$, and

$$\text{true (mean) reading} = \frac{300 \cdot m}{r} = \frac{300\,(r + m' - 300)}{r}$$

$= $ the number of minutes and seconds to be added to the degrees and minutes of the limb.

XVI. *Lunar distances.*

To determine the *true* distance of the moon from the sun, or a star; the *apparent* distance, together with the *apparent* altitudes of the moon and the sun, or star, being given.

Let,

d = apparent distance d' = true distance

H = moon's app't altitude H' = moon's true altitude

h = sun's app't altitude h' = sun's true altitude

P = moon's hor. par. at place P' = moon's par. in altitude

p = sun's hor. parallax p' = sun's par. in altitude

S = moon's hor. semidiam- S' = moon's augm. semidi-
     eter   ameter

s = sun's semidiameter D = observed distance

R = refraction for moon's r = refraction for sun's al-
     altitude   titude

A = observed altitude of a = observed altitude of
     moon's limb   sun's limb.

$P = \pi - \pi \, . \, E. \sin^2 L$; where π = moon's equatorial horizontal parallax.

E = the elipticity, $\log E = 7.5233789$; L = the latitude of place.

$$S = [9.43537] \, P \qquad S' = S + \text{augmentation}$$
$$H = A \pm S' \qquad h = a \pm s$$
$$P' = P \cos H \qquad p' = p \cos h$$
$$H' = H + (P' - R) \qquad h' = h - (r - p')$$

For a Star or a **Planet.**

$$h' = h - r \qquad\qquad d = D \pm S'$$

$$\sin^2 C = \frac{\cos h' \cos H'}{\cos h \cos H} \cos \tfrac{1}{2}(h + H + d)\, \cos \tfrac{1}{2}(h + H - d)$$

$$\sin^2 \tfrac{1}{2} d' = \cos \big\{\, \tfrac{1}{2}(h' + H') + C \,\big\} \cos \big\{\, \tfrac{1}{2}(h' + H') - C \,\big\}$$

The *reduced* distance being thus found, the longitude may be deduced from it as follows:

Suppose that at $5^{hrs}\ 05^m\ 56^s$ mean time, 29th April, 1838, at a place whose longitude is presumed to be $4^{hrs}\ 45^m\ 00^s$ west of Greenwich, the result of observations gave the *reduced* distance between the sun and moon, $d' = 71°\ 05'\ 35''$

Mean time obs'n $= 5^{hrs}\ 05^m\ 56^s$
Approx. long'de $= 4\quad 45\quad 00$
$$\overline{\qquad\qquad\qquad\qquad}$$
$\qquad\qquad\qquad\quad 9\quad 50\quad 56$ approx. Greenwich m.
$\qquad\qquad\qquad\qquad\qquad\qquad\qquad$ time of observation.

By Naut. Alm. at $IX^h = 70°\ 41'\ 30'' \qquad\qquad 70°\ 41'\ 30''$
(April 29th) $\qquad XII^h = 72°\ 07'\ 47'' \qquad d' = 71°\ 05'\ 35''$
$$\qquad\qquad\overline{\qquad\qquad\qquad}\qquad\qquad\overline{\qquad\qquad}$$
$\qquad\qquad\qquad\qquad\quad 1\quad 26\quad 17 \qquad\qquad\qquad\quad 24'\ 05''$

Increase of distance in $3^{hrs} = 5177''.0 \quad\ \delta\, d' = 1445''$
Then $5177'' : 10800^s :: 1445'' :\ x = 0^h\ 50^m\ 14^s.5$
$\qquad\qquad\qquad\qquad\qquad\qquad$ Add $\quad 9^{hrs}$
$$\qquad\qquad\qquad\qquad\qquad\qquad\overline{\qquad\qquad\qquad}$$
Greenwich mean time deduced $= 9^{hrs}\ 50^m\ 14^s.5$
Mean time at place $\qquad\qquad = 5\quad 05\quad 56.0$
$$\qquad\qquad\qquad\qquad\qquad\qquad\overline{\qquad\qquad\qquad}$$
Longitude, deduced $\qquad\qquad = 4^h\ 44^m\ 18^s.5$

The reduction of this proportion is very much facilitated by the use of *Proportional Logarithms,* or logs. of $\dfrac{3^{hrs}}{T}$ given in treatises on Navigation, in conjunction with those in the Nautical Almanac.

The proportion, however, requires a correction for second differences, when greater accuracy is desired, arising from the irregularity of the moon's motion.

A closer approximation to the true value of the quantity x being

$$X = \frac{3^{hrs} \, \delta \, d'}{A + \frac{1}{3} B \, x}$$

In which $B = \frac{1}{4}$ the sum of the second differences, and $A =$ the middle first difference $- B$; thus,

From the Nautical Almanac, April 29, 1838.

	1st difference.	2d difference.
At VI $= 69°$ 14$'$ 54$''$		
	$+ 1°$ 26$'$ 36$''$	
IX $= 70°$ 41$'$ 30$''$		$- 0'$ 19$''$
	$+ 1°$ 26$'$ 17$'' = \triangle_1$	
XII $= 72°$ 07$'$ 47$''$		$- 0'$ 18$''$
	$+ 1°$ 25$'$ 59$''$	
XV $= 73°$ 33$'$ 46$''$		

$x = 50^m \, 14^s.5 = 0^{hrs} \, 83736$ (table page 173.)

$B = - 9''.2$; $A = \triangle_1 - B = 5177'' + 9''.2 = 5186''.2$

$\delta d = 1445''$; $\frac{1}{3} B \, x = - 2''.56$; $A + \frac{1}{3} B \, x = 5183''.64$

whence

$$X = \frac{10800^s \times 1445''}{5183''.64} = 50^m \, 10^s.6.$$

and, longitude deduced $= 4^{hrs} \, 44^m \, 14^s.6.$

Reduction of the Moon's Equatorial Horizontal Parallax to the Horizontal Parallax in any Latitude.

LATITUDE.	HORIZONTAL PARALLAX.				
	54′	56′	58′	60′	62′
°	″	″	″	″	″
0	0.0	0.0	0.0	0.0	0.0
8	0.2	0.2	0.2	0.2	0.2
16	0.8	0.8	0.9	0.9	0.9
20	1.3	1.3	1.4	1.4	1.5
24	1.8	1.9	1.9	2.0	2.0
28	2.4	2.5	2.6	2.6	2.7
32	3.0	3.1	3.3	3.4	3.5
36	3.7	3.9	4.0	4.1	4.3
40	4.5	4.6	4.8	5.0	5.1
44	5.2	5.4	5.6	5.8	6.0
48	6.0	6.2	6.4	6.6	6.8
52	6.7	7.0	7.2	7.4	7.6
56	7.4	7.7	8.0	8.2	8.5
60	8.1	8.4	8.7	9.0	9.3
64	8.8	9.1	9.4	9.7	10.0
68	9.3	9.6	10.0	10.3	10.6
72	9.8	10.1	10.4	10.8	11.2
76	10.2	10.6	10.9	11.3	11.7
84	10.7	11.1	11.5	11.9	12.0
90	10.8	11.2	11.6	12.0	12.4

The moon's horizontal parallax, given in the second page of each month, in the "American Nautical Almanac," for noon and midnight, is the equatorial parallax for Greenwich mean noon and midnight; from thence it is to be deduced for the time and place of observation. The correction for latitude, on account of the spheroidal figure of the earth, can be made from the table above. Thus, supposing the hor. equat. par. to be 58′; the hor. par. in lat 52° would be 58′ — 7″.2 = 57′ 52″.8.

Augmentation of the Moon's Semidiameter, on account of her apparent altitude.

APPARENT ALTITUDE.	HORIZONTAL SEMIDIAMETER.					
	14′ 30″	15′ 0″	15′ 30″	16′ 0″	16′ 30″	17′ 0″
○	″	″	″	″	″	″
0	0.00	0.00	0.00	0.00	0.00	0.00
3	0.71	0.75	0.80	0.86	0.92	0.97
6	1.41	1.50	1.60	1.71	1.83	1.94
9	2.11	2.25	2.40	2.56	2.73	2.90
12	2.81	3.00	3.20	3.41	3.63	3.86
15	3.50	3.74	3.99	4.25	4.52	4.80
18	4.17	4.46	4.76	5.07	5.39	5.73
21	4.84	5.18	5.52	5.89	6.26	6.65
24	5.49	5.88	6.27	6.68	7.11	7.54
27	6.13	6.56	7.00	7.46	7.93	8.42
30	6.75	7.23	7.71	8.22	8.74	9.28
33	7.35	7.88	8.40	8.96	9.52	10.12
36	7.93	8.50	9.07	9.67	10.28	10.92
39	8.49	9.10	9.72	10.36	11.02	11.66
42	9.03	9.68	10.34	11.02	11.72	12.44
45	9.55	10.23	10.93	11.65	12.39	13.15
48	10.05	10.76	11.49	12.25	13.03	13.83
51	10.52	11.26	12.02	12.81	13.63	14.46
54	10.95	11.72	12.52	13.34	14.19	15.06
57	11.35	12.15	12.98	13.83	14.72	15.62
60	11.72	12.55	13.40	14.29	15.20	16.13
63	12.06	12.91	13.79	14.70	15.64	16.60
66	12.37	13.24	14.14	15.08	16.04	17.03
69	12.64	13.53	14.46	15.41	16.39	17.40
72	12.88	13.79	14.73	15.70	16.70	17.73
75	13.08	14.01	14.96	15.95	16.96	18.01
78	13.24	14.18	15.15	16.15	17.18	18.24
81	13.37	14.32	15.30	16.31	17.35	18.42
84	13.46	14.42	15.41	16.42	17.47	18.55
87	13.52	14.48	15.47	16.49	17.54	18.62
90	13.54	14.50	15.49	16.51	17.57	18.65

XVII. *Longitude by Lunar Culminations.*

1. *Interpolation.*—When the quantities in the ephemeris are given in intervals of 12^{hrs}, and the assumed meridian is $+$, or west of Greenwich, the following arrangement will be found convenient:

Let $a_1 =$ the moon's place, from the ephemeris, for the preceding noon or midnight,

$\quad a^1 =$ the moon's place, for the following midnight or noon,

$\quad a = \frac{1}{2}(a_1 + a^1)$,

$\quad b =$ the middle first difference,

$\quad c =$ the mean of the two middle second differences,

$\quad = \frac{1}{2}(c_1 + c^1)$,

$\quad d =$ the middle third difference,

$\quad e =$ the mean of the two fourth differences

$\quad = \frac{1}{2}(e_1 + e^1)$,

$\quad f =$ the fifth differences,

$\quad t =$ the interval in seconds since the date for a_1,

$\quad m =$ the variation of the moon's place for the interval $(t - 6^{hrs})$,

$\quad n =$ the average hourly variation,

$\quad n' =$ the true hourly variation at the instant, t,

Enclosing in brackets the constant log. co-efficients,

$$\text{Let } X = [5.3645163]\,(t - 6^h\ 0^m\ 0^s)$$
$$X' = [0.42800]\,t\,(t - 12^h\ 0^m\ 0^s)$$
$$X'' = [9.5229]\,XX'.$$
$$X''' = [9.6499]\,X'\,(t + 12^h\ 0^m\ 0^s)(t - 24^h\ 0^m\ 0^s)$$

Then:

$$m = b\,X + c\,X' + d\,X'' + e\,X'''$$

$$n = [\,3.25527\,]\left(\frac{b + 2\,m}{t}\right)$$

$$n' = \tfrac{1}{12}\left\{\,(b + \tfrac{1}{12}\,d - \tfrac{1}{120}\,f) + (c - \tfrac{1}{6}\,e)\,X + d\,X'\,\right\}$$

If the corrections beyond the *second* differences are neglected, then

$$m = b\,X + c\,X'$$

$$n = [3.25527]\left(\frac{b + 2\,m}{t}\right)$$

This will require but four quantities from the ephemeris; two preceding and two following the time t.

2. To apply this to *moon culminations*:

Let AR = the right ascension in the Nautical Almanac of the moon's bright limb at Greenwich for the upper transit next preceding the transit observed.

AR$'$ = the observed AR of the moon's bright limb at the place whose longitude, L, is required.

Assume an approximate longitude, l, and compute from the Nautical Almanac, by means of the foregoing method of interpolation, what the increase, m, in the AR of the moon's bright limb from its last transit at Greenwich, should be for that longitude.

As the correction, in this case, is to be applied to a_1, instead of $\frac{1}{2}\,(a_1 + a^1)$, the co-efficient X becomes $= [5.3645163]\,l$; the other co-efficients remaining the same, merely changing t to l.

By your own observations and the Nautical Almanac, this change is (AR$'$ — AR); then

$$\text{as } m : (\text{AR}' - \text{AR}) :: l : \text{L and}$$

$$\text{L} = (\text{AR}' - \text{AR})\cdot\frac{l}{m}$$

As the moon's motion in right ascension is not uniform, this proportion is only true when the values of l and L are nearly equal; but it is supposed that the approximate longitude is known to within a minute of time. It has also been supposed that both AR and AR′ are correctly determined; that the quantities in the Nautical Almanac are free from errors in the Lunar Tables, and those of nutation, &c.; and that the observed AR is corrected for error of clock and errors of position, etc., in the Transit instrument.

It is to eliminate these sources of error that *moon culminating stars* are observed in conjunction with the moon, and that corresponding observations at points whose positions are accurately known, are substituted for the tabular values, although the elements of the Nautical Almanac give very good approximations.

The complete method will be better illustrated by an example:

Suppose that at a station, whose longitude is presumed to be $4^h 55^m 50^s$ west of Greenwich, the following transits have been taken with a Chronometer marking sidereal time; the error of the Chronometer being immaterial, but the Transit instrument being supposed in the meridian, or very nearly so.

February 18,	ζ	Geminorum	6^h 54^m $41^s.75$				
	♂	Geminorum	7 10 38.97				
	☽'s 1st Limb	– – –			7^h 28^m $06^s.76$	
	ζ	Cancri	8 03 06.11				
		Sum	3) 22 08 26.83		7 22 48.943	
				Diff.	0 15 17.817	
Rate of Chronometer + 3^s, daily,					— .0318	
Corrected difference = t,					= 0 15 17.785	

And suppose the following to be the corresponding observations at Greenwich, (these, however, are from the Nautical Almanac.)

February 18, ζ Geminorum   6^h 54^m 57^s.41

δ Geminorum   7 10 54.36

$\mathcal{D}$'s 1st Limb   – – –   7^h 27^m 47^s.66

ζ Cancri   8 03 21.44

3) 22 09 13.21   7 23 04.403

Difference $= t'$ $= 0$ 04 43.257

$t_{,}$ $= 0$ 15 17.785

then $t_{,} - t' =$ observed increase in $\mathcal{D}$'s AR $= m' =$ 10 34.528

In the same manner would be obtained, for other corresponding observations, values of m'', m''', &c.

Next, compute this increase from the Nautical Almanac, as follows :

Approximate Longitude $= l = 4^h 55^m 50^s.$

$l = 4^h 55^m 50^s = 17750^s$; log $= 4.2491984$

constant . . . . . . $= 5.3645163$

log X . . . . . $= + 9.6137147$

$l - 12$ hrs. $= - 25450^s$; log $= - 4.40568$

log l . . . . $= + 4.24919$

constant . . . . . . $= + 0.42800$

log X' . . . . . . $= - 9.08287$

log X . . . . . . $= + 9.6137$

log X' . . . . . . $= - 9.0828$

constant . . . . . . $= + 9.5229$

log X'' . . . . . . $= - 8.2194$

$l + 12$ hrs. $= 60950^s$; log $= + 4.7849$

$l - 24$ hrs. $= 68650^s$; log $= - 4.8366$

log X' . . . . . . -9.0828

constant . . . . . . $+ 9.6499$

log X''' . . . . . . $+ 8.3542$

Day.	Culm.	AR ☽'s 1st limb by Naut. Almanac.	Δ_1	Δ_2	Δ_3	Δ_4
17	U	6^h 35^m 55^s.73				
			+ 26^m 00^s.54			
"	L	7 01 56.57				
			+ 25 51.39	— 0^m 09^s.15		
18	U	7 27 47.66	b=+ 25 41.18	c_I =— 0 10.21	— 01^s.06	e_l= + 0^s.81
"	L	7 53 28.84		c^1 =— 0 10.46	d =— 0.25	e_l= + 0.84
			+ 25 30.72		+ 0.59	
19	U	8 18 59.56		— 0 09.87		
			+ 25 20.85			
"	L	8 44 20.41				
Sum of differences . . .			2^h 08^m 24^s.68	— 0^m 39^s.69	— 0^s.72	+ 1.65
Upper left hand quantity .			6 35 55.73	26 00.54	— 0 9.15	— 1.06
Check. Sum + upper left / =lower left			8 44 20.41	25 20.85	— 0 9.87	+ 0.59
b, c, d, e, . . .			+ 1541^s.18	— 10^s.33	— 0^s.25	+ 0^s.82
log b, log c, log d, log e .			3.1878533	— 1.01431	— 9.3979	+ 9.9138
log X, log X$'$, log X$''$, log X$'''$,			+ 9.6137147	— 9.08187	— 8.2194	+ 8.3542
log b X, log c X$'$, etc . .			+ 2.8015680	+ 0.09718	+ 7.6173	+ 8.2680
b X, c X$'$, d X$''$, etc . .			+ 633^s.240	+ 1^s.250	+ 0^s.004	+ 0^s.018

$$
\begin{aligned}
b\,\mathrm{X} &= 633^s.240 \\
c\,\mathrm{X}' &= + 1.250 \\
d\,\mathrm{X}'' &= + .004 \\
e\,\mathrm{X}''' &= + .018 \\
\hline
634^s.512 &= m
\end{aligned}
$$

$$
\begin{aligned}
\log l &= 4.2491984 \\
\log m &= 2.8024399 \\
\hline
\log \frac{l}{m} &= 1.4467585
\end{aligned}
$$

Then,

    observed increase $= 634.528 = m'$

    computed   do. $= 634.512 = m$

    observed excess $= m' - m = 0^s.016.$

$$\text{Longitude, deduced,} = 4^h\ 55^m\ 50^s + \frac{l}{m}(m' - m)$$

$$= 4_h\ 55^m\ 50^s.45$$

If there are corresponding observations at some other well known point, say Cambridge, Mass., longitude $= 4^h\ 44^m\ 32^s = l'$; compute the increase m_l for this longitude, by changing the co-efficients X, X$'$, X$''$, etc., to correspond to l'. Then $(m - m_l)$ will be the computed increase for Cambridge to your station, and $\frac{l'}{m_l}$ the rate of this increase, with which proceed as above.

It often happens that two observers do not use the same number of wires, or that the same number of stars are not observed at the two places. In such cases the observed increase of the right ascension of the moon's limb requires a correction, which Mr. Walker deduces as follows, from Gauss's method :

For the European observatory and western station respectively,

Let A$'$ and A $=$ the observed AR of a star,

        E $=$ A$'$ $-$ A for the same star,

        E $=$ a similar value for another star,

    l and l' $=$ the number of wires on which each limb was observed,

    a and a' $=$ similar values for a star,

$$\lambda = \frac{l\,l'}{l + l'}, \text{ for the moon's limb,}$$

$$u = \frac{a\,a'}{a + a'}, \text{ for one star.}$$

$u' =$ a similar value for another star,

$\Sigma =$ symbol to denote the aggregate of similar quantities,

$\varepsilon =$ the correction required.

$$\text{Then } \varepsilon = \frac{\Sigma \left(\mathrm{E}\,\dfrac{\lambda\,u}{\lambda + u} \right)}{\Sigma\,\dfrac{\lambda\,u}{\lambda + u}}$$

$$\text{and } \mathrm{L} = l + \frac{l}{m}\,(m' - m + \varepsilon)$$

Also, calling W, the *weight* of each day's comparison,

$$\mathrm{W} = \frac{\sigma\,\lambda}{(\sigma + \lambda)\,z^2}$$

in which z is the same as $\dfrac{l}{m}$ and $\sigma = u + u' + u''$, etc.

For the weight of the result of all the comparisons, we have

$$\Sigma\,\mathrm{W} = \Sigma\,\frac{\sigma\,\lambda}{(\sigma + \lambda)\,z^2}$$

Let e denote the probable error of observation, and E the probable error of the final result; then,

$$\mathrm{E} = \Sigma\,\frac{e}{\sqrt{\Sigma\,\dfrac{\sigma\,\lambda}{(\sigma + \lambda)\,z^2}}}$$

It frequently happens that the moon cannot be observed on the middle wire, in which case she is far enough from the meridian to have a sensible parallax in right ascension; and as it may be very desirable not to lose the observation, this parallax must be computed and applied to the hour angle from the middle wire, which is supposed to be nearly coincident with the meridian.

Denoting this parallax in right ascension by p, the horizontal parallax by w, the latitude of the place of observation by ϕ, and the true declination of the moon by δ, we have from the ordinary series for the parallax in right ascension, neglecting the terms after the first, which would in this case be insignificant,

$$p = \theta \sin w \cos \phi \sec \delta,$$

in which θ, is the hour angle, or equatorial interval in sidereal time from the lateral wire on which the moon is observed to the central wire; so that, at the instant of observation, the actual distance of the moon's limb from the central wire is:

$$\theta - \theta \sin w \cos \phi \sec \delta,$$

and the reduction to meridian or middle wire will be

$$\pm \frac{\theta}{\cos \delta} \cdot \frac{1 - \sin w \cos \phi \sec \delta}{1 - 0.00277\, m}$$

in which m, is the motion of the moon in right ascension in one day, expressed in degrees. The upper sign is to be used when the observation is on a wire *before*, and the lower *after* the middle wire.

XVIII. *The value of a quantity at three consecutive whole hours, T—1, T and T + 1, being given, to find its value at an intermediate time T′, and its hourly variation at that time.*

Attending to the algebraic signs, subtract the value of the quantity at the time T — 1, from its value at the time T; and its value at the time T, from its value at the time T + 1; and the remainders will be the *first differences.* Subtract the first of these from the second, and the remainder will be the *second difference.* Let $a =$ the value of the quantity at the time T; $b =$ the half sum of the first differences; $c =$ the second difference; and $t =$ the interval between T and T′, expressed in the fraction of an hour, and marked negative when T′ is *earlier* than T. Then the value of the quantity at the time T′, will be

$$a + t\, b + \frac{t^2}{2}\, c:$$

And the hourly variation of the quantity at the time T′, will be $\qquad b + t\, c.$

EXAMPLE.

Given the moon's declination, on a certain day, as follows:

At 10^h, D $= + 15^\circ 58' 50''.1$; at 11^h, D $= 15^\circ 47' 11''.0$;

At 12^h, D $= \quad 15^\circ 35' 27''.1$. Required its value at $10\frac{3}{5}{}^h$.

	D	1st differences.	2d difference.
10	$+ 15^\circ 58' 50''.1$		
		$- 11' 39''.1$	
11	$15^\circ 47' 11''.0$		$- 4''.8$
		$- 11' 43''.9$	
12	$15^\circ 35' 27''.1$		

$a = + 15^\circ 47' 11''.0,\; b = - 11' 41''.5,\; c = - 4''.8,\; t = - \frac{2}{5}$

$t\, b = + \qquad 4' 40''.6$

$\dfrac{t^2}{2} c = - \qquad\qquad 0''.4$

D $= + 15^\circ 51' 51''.2$ at time T′

$\qquad\qquad\qquad b = - 11' 41''.5$

$\qquad\qquad\qquad t.\, c = + \qquad 1''.9$

Hourly variation at time T′ $= - 11' 39''.6$

XIX. *To find the Longitude of a place from an observed occultation of a fixed star by the Moon.*

Let A $=$ Moon's AR,

A$'$ $=$ Star's AR,

D $=$ Moon's declination,

D$'$ $=$ Star's declination,

A$''$ $=$ Moon's hourly variation in AR,

D$''$ $=$ Moon's hourly variation in declination,

π $=$ Moon's equatorial horizontal parallax,

H$'$ $=$ Star's hour angle for Greenwich,

k $= \dfrac{\text{sine moon's appt. semidiam.}}{\pi} =$ constant $= 0.2725,$

$$\log = 9.43536,$$

ϕ $=$ Geographical north latitude of place,

ϕ' $=$ Geocentric north latitude of place,

ρ $=$ Earth's radius at place.

It is unnecessary to compute ϕ' and ρ separately, as

$$\rho \sin \phi' = \frac{(1-e^2)}{\sqrt{1-e^2 \sin^2 \phi}} \cdot \sin \phi = A \sin \phi$$

$$\rho \cos \phi' = \frac{1}{\sqrt{1-e^2 \sin^2 \phi}} \cdot \cos \phi = B \cos \phi$$

in which $e = .081697 =$ the Earth's eccentricity; and as the values of log A and log B may be taken from the following table, with the argument ϕ :

ϕ	Log A.	Log B.
0°	9.9971	0.0000
10	9.9971	0.0000
20	9.9973	0.0002
30	9.9975	0.0004
40	9.9977	0.0006
50	9.9979	0.0009
60	9.9982	0.0011
70	9.9984	0.0013

1. With the estimated Longitude of the place, reduce the observed mean time of immersion to Greenwich time. Let T stand for this time, and T′ for the same time, taken to the nearest tenth of an hour. From the Nautical Almanac, find for the time T′ by the problem on page 232, the values of A, D, A″, D″, and by proportion, the value of π; and also take out the values of A′, D′, and the sidereal time of mean moon.

2. With the values of A, D, etc., at the time T′, find the values of p, q, $p′$ and $q′$, from the following formulæ:

$$p = \frac{(A-A') \cos D}{\pi}, \qquad c = \frac{D-D'}{\pi}$$

$$\log B = \log p + \log \sin D' + 4.6856$$

$$d = B\,(A-A'); \qquad q = c + \tfrac{1}{2}\,d,$$

$$a' = \frac{A'' \cos D}{\pi}, \qquad b' = \frac{D''}{\pi}$$

$$c' = B\,A'' \qquad d' = B\,D''$$

$$p' = a' - d' \qquad q' = b' + c'$$

3. To the sidereal time at mean noon, add the sidereal time corresponding to the interval that T is past noon, and from the sum subtract A′. To the remainder, apply the longitude of the place in time, by *adding* if it is *east*, but *subtracting* if it is *west*, and converting the result into degrees, it will be H, the star's hour angle at the observed time of immersion.

4. Having found log ρ cos ϕ' and log ρ sin ϕ', for the place, find u, v, N, F, t and t'' by the following formulæ:

$$f = \rho \sin \phi' \cos D'$$

$$u = \rho \cos \phi' \sin H, \qquad g = \rho \cos \phi' \cos H \sin D'$$

$$v = f - g ,$$

$$\cot N = \frac{q'}{p'} \quad , \qquad\qquad d = (p - u) \cot N ,$$

$$\cos F = \frac{(d + v - q) \sin N}{k} ,$$

$$t = \frac{k \cos (N + F)}{p'} , \qquad\qquad t'' = \frac{p - u}{p'} .$$

Then will $T' - t'' + t$, be the corrected Greenwich mean time of the immersion. The difference between this and the observed time, will be the Longitude in time; *west* if the observed time is the *earlier* of the two, but *east*, if it is *later.*

In a similar manner would be deduced the Longitude from the observed emersion, except, that instead of t, we find $t' = \dfrac{k \cos (N - F)}{p'}$. When the immersion and emersion have both been observed, the Longitude should be obtained from each, and the mean of the two results taken.

EXAMPLE.

Suppose the observed immersion of i Leonis, on Jan. 7th, 1836, at a place in Latitude 52° 08′ 28″ N., estimated Longitude 0 h. 1 m. W., was 10 h. 45 m. 53.3 sec., mean time; required the Longitude of the place.

The observed time of immersion reduced to Greenwich time is, $T = 10$ h. 46 m. 53.3 sec. Taking $T' = 10.8$ h. $= 10\frac{4}{5}$ h., we easily find from the Nautical Almanac,

$A = 10^h\ 20^m\ 33^s.89$ $D = +\ 15°\ 49'\ 31''.2$

$A' = 12\ \ 23\ \ 26.39$ $D' = +14°\ 58'\ 38''.8$

$A'' = 122^s.905 = $ (in arc), $1843''.6$; $D'' = - 700''.5$

$A - A' = - 2587''.5$; $D - D' = 3052''.4$; $\pi = 3362''.0$

$A - A'$ . . . . . log 3.41288 — $D - D' = $ . . . . log 3.48464

    $\pi = $ Ar. Co. " 6.47340 $\pi = $ Ar. Co. log 6.47340

    $D = $ . . . . cos 9.98322 $c = .9079$ $= 9.95804$

$p = - $ . . . $.7404 = 9.86950$ — $\frac{1}{2} d = .0012$

    $D' = $ . . . sin 9.4124 $q = .9091$

            4.6856

        $B = 3.9675$ —

$A - A'$ . . . . . log 3.4129

    $d = \ \ .0024\ = 7.3804$

     $A'' = $ . . . . log 3.26567 D'' . . . . . log 2.84541 —

π . . . Ar. Co. log 6.47340 π . Ar. Co. log 6.47340

D . . . . cos 9.98322 $b'' = - .2084 = 9.31881$ —

$a' = .5276 = $ . . . 9.72229

        $B = \ 3.9675$ — $B = 3.9675$ —

A'' . . . . 3.2657 — D'' . . . . . . 2.8454 —

$c' = - .0017 = \ \ 7.2332$ — $d' = .0006 = \ 6.8129$

$p' = a' - d' = .5270$; $q' = b' + c' = - .2101$

Sidereal time at mean noon Greenwich from N. A. $= 19^h\ 04^m\ 22^s.41$

Sidereal interval from noon to time T . . . . . $10\ \ 48\ \ 39.57$

                                     $5\ \ 53\ \ 01.98$

A' . . . . $10\ \ 23\ \ 26.39$

                              $-\ \ 4\ \ 30\ \ 24.41$

                                  $1\ \ 00$

            $H = - 67°\ 51'\ 06'' = -\ \ 4\ \ 31\ \ 24.41$

From page 233 we have

$$\log \varrho \cos \varphi' = 9.78888 \qquad\qquad \log \varrho \sin \varphi' = 9.89538$$

$$\begin{aligned}
\varrho \sin \varphi' &= 9.89538 \\
\text{D}' \cos\ & \underline{9.98499} \\
f = .7592 &= 9.88037
\end{aligned}$$

$\varrho \cos \varphi$ . . . 9.78888		$\varrho \cos \varphi$ . . 9.78888
H . . . sin $\overline{9.96671}$ —		H . . cos 9.57635
$u = -\ .5696\ =\ 9.75559$ —		D' . . sin $\overline{9.41236}$
$v =\quad .6993$		$g = .0599\ =\ 8.77759$

q' . . . . . log 9.32243 —   N . . sin 9.96797

p' Ar. Co.   log $\overline{0.27819}$ $d + v - q = -\ .1417$ log 9.15137

N = 111° 44′ 10″   cot 9.60062 — k Ar. Co. log 0.56463

$p - u = -\ .1708 = \log 9.23249$ —,   F = 118° 53′ 00″ cos 9.68397 —

$d =\quad .0681\ =\quad \overline{8.83311}$

N + F = 230° 37′ 10″ cos 9.80241 —, $p - u$ . . log 9.23249 —

k . . . . . log 9.43537 p' Ar. Co.   log 0.27819

p' Ar. Co.   log $\overline{0.27819}$ $t'' = -\quad .3241 = 9.51068$ —

$t = -\ .3281 \qquad 9.51597$ —

$$\begin{aligned}
\text{T}' - t'' + t \quad &= 10^\text{h}\ 47^\text{m}\ 45^\text{s}.6 \\
\text{Observed time} &= 10\quad 45\quad 53.3
\end{aligned}$$

$$\text{Longitude of place} = \qquad 1^\text{m}\ 52^\text{s}.3\ \text{west.}$$

XIX. *Formulæ for Probable Error and Precision.*

1. Let n, n', etc., = the results found by observation.

x = their arithmetical mean; or the result deduced from these by the method of least squares,

E_2 = (mean error),

E_1 = (mean error)^1 = arithmetical mean without regard to sign,

m = the number of observations,

r = probable error of a single observation, n, n', etc.,

R = probable error of final result, x,

h = measure of exactness of a single observation n, n',

H = measure of exactness of final result, x,

w = probable error of an observation assumed as the standard of excellence,

p, p', etc., = the weights of the determinations of several variables,

Σ = symbol representing the sum,

$$E_2 = \sqrt{\frac{\Sigma\,(x-n)^2}{m-1}} \qquad E_1 = \frac{\Sigma\,(x-n)}{m-1}$$

$$r = 0.674489\ E_2 \qquad r = 0.845347\ E_1$$

$$h = \frac{0.469360}{r} \qquad H = h\sqrt{m}$$

$$R = \frac{r}{\sqrt{m}}$$

2. By the *weight* of any determination, is meant its relative approximation to the true value.

It may be measured by the number of equally good observations (one of which is assumed to represent the unit of excellence) necessary to give a result equally near the true value.

The weights of two determinations are to each other in the direct proportion of the squares of their relative measures of exactness, and in the inverse proportion of the squares of their probable errors:

$$p = \frac{w^2}{r^2}, \qquad\qquad p' = \frac{w^2}{r'^2}$$

and calling the weight of any function of the two determinations, whose weights are p and p',

$$P = \frac{pp'}{p + p'}$$

the probable error of the value of the function is

$$R' = \frac{w}{P} = \sqrt{r'^2 + r^2}$$

If the index or measure of precision vary as any element v, involved in any given determination,

or $h : h' :: v : v' \ \therefore \ h' = \dfrac{h\,v'}{v}$, then will the weight become (see page 230.)

$$P = \frac{p\,p'\,v'^2}{p + p'},$$

If there be but a single variable, and this has been found by different examinations, giving the values of a, a', a'', etc., with probable errors r, r', r'', etc., or the weights p,

p', p'', etc.; and we seek to find from them the most probable value of x,

$$x = \frac{a\,p + a'\,p' + a''\,p'' + \text{etc.}}{p + p' + p'' + \text{etc.}} = \frac{\frac{a}{r^2} + \frac{a'}{r'^2} + \frac{a''}{r''^2} + \text{etc.}}{\frac{1}{r^2} + \frac{1}{r'^2} + \frac{1}{r''^2} + \text{etc.}};$$

Its weight

$$\mathrm{P} = p + p' + p'' + \text{etc.};$$

Its probable error

$$\mathrm{R}'' = \frac{1}{\sqrt{\dfrac{1}{r^2} + \dfrac{1}{r'^2} + \dfrac{1}{r''^2} + \text{etc.}}}$$

Position of some of the principal Observatories, etc.

PLACE.	Latitude North.	LONGITUDE				AUTHORITIES.
		From the Observatory at Washington.		From the Observatory at Greenwich.		
		In time.	In arc.	In time.	In arc.	
	° ′ ″	H M S	° ′ ″	H M S	° ′ ″	
Greenwich Observatory	51 28 38.2	E. 5 08 11.2	77 02 48.0	....	....	Am. Naut. Alm.,
Paris "	48 50 13.2	E. 5 17 32.7	79 23 09.9	E. 0 09 21.5	2 20 21.9	(1855)
Cambridge, Mass. "	42 22 48.6	E. 0 23 41.5	5 55 23.1	W. 4 44 29.6	71 07 24.9	"
Cincinnati, Ohio, "	39 05 54.0	W. 0 29 46.9	7 26 42.8	W. 5 37 58.0	84 29 30.8	"
Georgetown, D. C., "	38 54 26.1	W. 0 00 06.2	0 01 33.0	W. 5 08 17.4	77 04 21.0	"
Hudson, Ohio, "	41 14 42.6	W. 0 17 32.1	4 23 00.9	W. 5 25 43.3	81 25 48.9	"
Philadelphia "	39 57 07.5	E. 0 07 33.6	1 53 24.6	W. 5 00 37.6	75 09 23.4	"
Washington "	38 53 39.3	....	....	W. 5 08 11.2	77 02 48.0	"
Washington, Capitol.....	38 53 19.9	E. 0 00 10.2	0 02 33.0	W. 5 08 01.0	77 00 15.0	Coast Survey (1852)

31

Falls St. Anthony, U. S. Cottage,	Lat. = 44° 58′ 40″	Nicollet.
	Lon. = 6ʰ 12ᵐ 42ˢ	"
Fort Leavenworth, Landing . .	Lat. = 39° 31′ 14″	Emory.
	Lon. = 6ʰ 18ᵐ 56ˢ	Nicollet.
Council Bluffs	Lat. = 41° 25′ 04″	Graham.
	Lon. = 6ʰ 22ᵐ 55ˢ.5	"
Fort Gibson, old block house .	Lat. = 35° 47′ 34″.8	Woodruff.
	Lon. = 6ʰ 21ᵐ 00ˢ.9	"
San Antonio, Texas	Lat. = 29° 25′ 22″.0	Johnston.
	Lon. = 6ʰ 33ᵐ 57ˢ	"
Paso del Norte, Plaza . . .	Lat. = 31° 44′ 16″	Salazar.
	Lon. = 7ʰ 05ᵐ 15ˢ	"
Frontera, White's rancheria . .	Lat. = 31° 48′ 39″	Whipple.
	Lon. = 7ʰ 05ᵐ 54ˢ	"
Santa Fé	Lat. = 35° 41′ 06″	Emory.
	Lon. = 7ʰ 04ᵐ 10ˢ	"
Bent's Fort	Lat. = 38° 02′ 22″	Fremont.
	Lon. = 6ʰ 54ᵐ 13ˢ.3	"
Fort Laramie	Lat. = 42° 12′ 10″	Fremont.
	Lon. = 6ʰ 59ᵐ 10ˢ.9	"
Fort Hall	Lat. = 43° 01′ 30″	Fremont.
	Lon. = 7ʰ 29ᵐ 59ˢ.6	"
San Diego, Coast Survey obs'y,	Lat. = 32° 41′ 57″.9	Coast Sur'y,
	Lon. = 7ʰ 48ᵐ 53ˢ.4	Rep't of '51.
Point Conception C. S. obs'y, .	Lat. = 34° 26′ 56″.3	Coast Sur'y,
	Lon. = 8ʰ 01ᵐ 42ˢ.2	Rep't of '51.
Point Pinos, Coast Survey obs'y	Lat. = 36° 37′ 59″.8	Coast Sur'y,
	Lon. = 8ʰ 07ᵐ 37ˢ.4	Rep't of '51.
San Francisco, Presidio Hill .	Lat. = 37° 47′ 35″.6	Coast Sur'y,
	Lon. = 8ʰ 09ᵐ 47ˢ.2	Rep't of '51.

Longitudes west from Greenwich.